Glycals and their Derivatives

Edited by

Nazar Hussain

Department of Medicinal Chemistry
Institute of Medical Sciences
Banaras Hindu University, Varanasi-221005
India

&

Atul Kumar

Department of Chemistry
Indian Institute of Technology
Bombay, Mumbai-400076, India

Glycals and their Derivatives

Editors: Nazar Hussain & Atul Kumar

ISBN (Online): 978-981-5322-79-8

ISBN (Print): 978-981-5322-80-4

ISBN (Paperback): 978-981-5322-81-1

need for a court order if at any point you breach any terms of this License Agreement. In no event will any delay or failure by Bentham Science Publishers in enforcing your compliance with this License Agreement constitute a waiver of any of its rights.

3. You acknowledge that you have read this License Agreement, and agree to be bound by its terms and conditions. To the extent that any other terms and conditions presented on any website of Bentham Science Publishers conflict with, or are inconsistent with, the terms and conditions set out in this License Agreement, you acknowledge that the terms and conditions set out in this License Agreement shall prevail.

Bentham Science Publishers Pte. Ltd.
80 Robinson Road #02-00
Singapore 068898
Singapore
Email: subscriptions@benthamscience.net

BENTHAM SCIENCE

CONTENTS

FOREWORD i

PREFACE ii

LIST OF CONTRIBUTORS iii

CHAPTER 1 INTRODUCTION TO GLYCALS 1
Bindu Tiwari, Mittali Maheshwari, Altaf Hussain, Ram Pratap Pandey and *Nazar Hussain*
   INTRODUCTION 1
   PREPARATION OF SIMPLE GLYCALS 4
   SYNTHESIS OF GLYCALS FROM GLYCOSYL SULFOXIDES 6
      Preparation of Pyranoid Glycals 6
      Preparation of Furanoid Glycals 7
   SYNTHESIS OF GLYCALS FROM THIOGLYCOSIDES OR GLYCOSYL SULFONES 9
   SYNTHESIS OF GLYCALS BY ELECTROCHEMICAL METHOD 10
   SYNTHESIS OF GLYCALS USING TITANIUM(III) REAGENT 12
   TRANSFORMATIONS OF GLYCALS 13
   GLYCALS IN SYNTHETIC AND NATURAL PRODUCTS 14
   EXO-GLYCALS 14
      Synthesis of Exo-glycals from Sugar Lactones 15
      Synthesis and Chemoselective Oxidation of Thiomethylphosphono-exo-glycals 19
      Synthesis of Difluoro Exo-glycals 20
   CONCLUSION 21
   REFERENCES 21

CHAPTER 2 RECENT ADVANCES IN THE SYNTHESIS OF O- AND S-GLYCOSIDES FROM GLYCALS 24
Mittali Maheshwari, Bindu Tiwari, Manish Kumar Sharma and *Nazar Hussain*
   INTRODUCTION TO O-GLYCOSIDES 24
   HISTORICAL ASPECTS 26
      Medicinal Applications of O-glycosides 27
      Mechanistic Insight into O-glycosylation 27
      Factors Affecting the Stereochemistry of the Glycosides 27
         Neighboring Group Participation 29
         Anomeric Effect 29
         Solvent Effect 30
      Literature on Synthesizing O-glycosides 30
   INTRODUCTION TO S-GLYCOSIDES 57
      Medicinal Applications of S-glycosides 58
      Literature On Synthesizing S-glycosides 59
   CONCLUSION 61
   REFERENCES 63

CHAPTER 3 RECENT ADVANCES IN THE SYNTHESIS OF C-GLYCOSIDES USING GLYCALS AND THEIR DERIVATIVES 69
Manish Kumar Sharma, Anand Kumar Pandey, Ram Pratap Pandey and *Nazar Hussain*
   SYNTHESIS OF C-GLYCOSIDES USING GLYCALS AND THEIR DERIVATIVES 72
      Ferrier Type C-glycosylation 72
      Transition-metal (TM)-catalyzed C-glycosylation 74
         Palladium-catalyzed C-glycosylation 76

       Cobalt-catalyzed C-glycosylation 88

       Nickel-catalyzed C-glycosylation 92

       Iridium-Catalyzed C-glycosylation 95

    Other Types of C-glycosylation 97

  CONCLUSION 102

  REFERENCES 102

CHAPTER 4 EXPLORING THE C-2 POSITION OF GLYCALS: STRUCTURAL INSIGHTS AND SYNTHETIC APPLICATIONS 106

  Ram Pratap Pandey, Manish Kumar Sharma, Anand Kumar Pandey, Altaf Hussain and Nazar Hussain

  INTRODUCTION 106

    2-Haloglycals 108

    C-H Activation 116

    1,2-Annulated Sugars 122

    2-Nitroglycals 125

    Cyclopropanated Sugars 127

  CONCLUSION 132

  ABBREVIATIONS 133

  REFERENCES 133

CHAPTER 5 TOTAL SYNTHESIS OF NATURAL PRODUCTS AND MEDICINALLY IMPORTANT MOLECULES FROM GLYCALS 137

  Norein Sakander and Qazi Naveed Ahmed

  INTRODUCTION 137

    Synthesis of Diospongin A 138

    Synthesis of Bradyrhizose 140

    Synthesis of Bergenin 140

    Total synthesis of (+)-aspicilin 141

    Synthesis of Reblastatin 143

    Synthesis of Cyclopropane Derivative of Spliceostatin A 145

    Synthesis of Papulacandins A-D and its Derivatives 146

    Synthesis of (−)-Hyacinthacine A1 149

    Synthesis of Tricyclic flavonoid 150

    Sporiolide B and its derivatives 150

    Synthesis of Cryptopyranmoscatone 152

    Synthesis of Biologically Potent Molecule: Tetrahydroquinolines 153

    Synthesis of Conduramine F4 154

    Synthesis of Aspergilide A 155

    Synthesis of Vineomycinone B2 157

    Synthesis of Oxadecalin Core of Phomactin A 157

    Synthesis of Peptidomimetics 159

    Synthesis of (−)-steviamine 160

    Synthesis of decytospolide A and B 162

    Synthesis of Spliceostatin G 163

  CONCLUSION 164

  ABBREVIATIONS 164

  REFERENCES 166

SUBJECT INDEX 174

FOREWORD

Carbohydrates are fascinating molecules that constitute an important class of naturally occurring compounds. They are most abundant in nature, biocompatible with minimal toxicity, and structurally diverse. Because of their key role in almost all biological processes, for example, cellular interaction, intercellular adhesion, signal transduction, inflammation, immune response, metastasis, fertilization, transport, modulation of protein function, cell surface recognition, and many more, there is an increased demand of carbohydrate-based molecules for their complete chemical, biochemical, and pharmacological investigations. For a long time, they have been explored as essential components in multiple vaccines and pharmaceuticals.

Among various carbohydrate entities, glycals (unsaturated sugars) and their derivatives particularly exhibit promising biological activities, ranging from antimicrobial to anticancer effects. The book "**Glycals and their Derivatives**" is timely, appropriate, and relies on carbohydrate-derived unsaturated sugars in drug discovery and development. The chapters in this book deliberate the practical utilities of a wide spectrum of diverse glycosides and 2-*C*-branched sugars by exploring their structural insights. This book provides a comprehensive overview of the different synthetic strategies for synthesizing crucial sugar analogs such as glycohybrids and glycoconjugates derived from glycals, offering a new avenue of understanding to the reader regarding glycoscience, glycobiology, and glycotechnology. The chemistry of glycals and their derivatives encompasses a vast array of reactions and transformations. Their reactivity is largely governed by the presence of the electron-rich double bond, which renders them susceptible to various electrophilic additions and cycloadditions. This reactivity is harnessed in the synthesis of diverse carbohydrate structures, including oligosaccharides and *O-, N-, S-,* and *C*-glycosides, which are pivotal in numerous biological functions and have profound implications for drug design and development.

This book, "*Glycals and their Derivatives*", also delves into the pharmacological profiles, selectivity, and metabolic stability of the intended sugar derivatives by exploring their post-modifications. Overall, this book has made a fantastic effort to acquaint the reader with the scope and emerging applications of carbohydrate-derived unsaturated sugars. I heartily congratulate the editors, Dr. Nazar Hussain and Dr. Atul Kumar, on making an eccentric and fruitful effort with this forthcoming valuable book, which certainly is a promising tool for synthetic glycochemists and glycobiologists.

Vinod Kumar Tiwari
Department of Chemistry
Institute of Science
Banaras Hindu University
Varanasi-221005, India

PREFACE

In the vast landscape of organic chemistry, the pursuit of novel molecules and functional groups has been a driving force behind groundbreaking discoveries and technological advancements. Among these, glycals (unsaturated sugars) and their derivatives stand as remarkable entities, captivating the interest of chemists across diverse fields. These cyclic enolic ethers are recognized as versatile chiral building blocks and are employed as starting materials for the total synthesis of various biologically active molecules. They have been extensively utilized in numerous glycosylation reactions, especially for the synthesis of deoxy sugars, Ferrier products, oligosaccharides synthesis, and cross-coupling reactions in the synthesis of C2-branched sugars. Originating from carbohydrates, the fundamental building blocks of life, glycals offer a fascinating intersection of biology and chemistry, promising avenues for both basic research and practical applications.

The unique reactivity of glycals lies in the presence of endocyclic and exocyclic double bonds, which offer multiple avenues for their transformation into various biologically important building blocks. In the realm of medicinal chemistry, glycals and their derivatives exhibit promising biological activities, ranging from antimicrobial and antiviral properties to anticancer effects. Structural modifications enable fine-tuning of their pharmacological profiles, enhancing potency, selectivity, and metabolic stability. Additionally, glycals have emerged as potential therapeutic agents for the treatment of diseases such as diabetes and inflammation, owing to their interactions with carbohydrate-processing enzymes and receptors.

The proposed book, "Glycals and their Derivatives", presents an exceptional compilation of cutting-edge research on carbohydrate-derived unsaturated sugars. It embodies the most recent scientific breakthroughs in organic and medicinal chemistry, offering an in-depth examination of glycals and their extensive derivatives. With its abundance of insights and discoveries, this book is positioned to become an essential reference for professionals, students, researchers, and academics involved in the dynamic realms of drug discovery and development.

Nazar Hussain
Department of Medicinal Chemistry
Institute of Medical Sciences
Banaras Hindu University, Varanasi-221005
India

&

Atul Kumar
Department of Chemistry
Indian Institute of Technology
Bombay, Mumbai-400076, India

List of Contributors

Altaf Hussain	Government Degree College, Budhal, J&K Higher Education Department, Jammu and Kashmir-185233, India
Bindu Tiwari	Department of Medicinal Chemistry, Institute of Medical Sciences, Banaras Hindu University, Varanasi-221005, India
Anand Kumar Pandey	Department of Medicinal Chemistry, Institute of Medical Sciences, Banaras Hindu University, Varanasi-221005, India
Manish Kumar Sharma	Department of Medicinal Chemistry, Institute of Medical Sciences, Banaras Hindu University, Varanasi-221005, India
Mittali Maheshwari	Department of Medicinal Chemistry, Institute of Medical Sciences, Banaras Hindu University, Varanasi-221005, India
Norein Sakander	Natural Products and Medicinal Chemistry Division, CSIR-Indian Institute of Integrative Medicine, Canal Road, Jammu-180001, India
Nazar Hussain	Department of Medicinal Chemistry, Institute of Medical Sciences, Banaras Hindu University, Varanasi-221005, India
Qazi Naveed Ahmed	Academy of Scientific and Innovative Research (AcSIR), Ghaziabad-201002, India
Ram Pratap Pandey	Department of Medicinal Chemistry, Institute of Medical Sciences, Banaras Hindu University, Varanasi-221005, India

CHAPTER 1

Introduction to Glycals

Bindu Tiwari[1], Mittali Maheshwari[1], Altaf Hussain[2], Ram Pratap Pandey[1] and Nazar Hussain[1,*]

[1] *Department of Medicinal Chemistry, Institute of Medical Sciences, Banaras Hindu University, Varanasi-221005, India*

[2] *Government Degree College, Budhal, J&K Higher Education Department, Jammu and Kashmir-185233, India*

Abstract: Glycals are 1,2-unsaturated sugars having a C=C bond between C-1 and C-2 of the pyranose or furanose moieties of the carbohydrate scaffold. The presence of a C=C bond leads to enhanced reactivity. They are used as chiral building blocks in the synthesis of various natural products and medicinally significant molecules. There are numerous methods for the synthesis of exo-glycals and endo-glycals, such as glycosidations, reagent-based methods, and electrochemical methods. Glycals have shown their versatility and applicability in chemical synthesis in many ways *e.g.*, epoxidation, cycloaddition, and formation of various glycosides like *C*-glycosides, *O*-glycosides, *N*-glycosides. They have also been employed in the synthesis of many biologically relevant natural products as starting materials.

Keywords: Exo-glycals, Furonoid glycals, Glycals, Reactivity, Synthesis.

INTRODUCTION

Carbohydrates are polyhydroxy aldehydes or ketones. They are firstly produced by plants and form a very large group of naturally occurring products. Carbohydrates are primarily used as biosynthetic precursors and structural elements in all living organisms. They exist as organic molecules in nature; some are used as energy suppliers and some as storage vehicles. They exist in simple forms as monosaccharides and disaccharides. They also exist as more complex glycosides like glycolipids, glycoproteins, peptidoglycans, proteoglycans, nucleic acids, and poly and lipopolysaccharides [1, 2]. Emil Fischer (1852-1990) (Fig. **1**) is regarded as the father of carbohydrate chemistry. He was one of the pioneering scientists in the area of organic chemistry.

---
[*] **Corresponding author Nazar Hussain:** Department of Medicinal Chemistry, Institute of Medical Sciences, Banaras Hindu University, Varanasi-221005, India; E-mail: nazar10@bhu.ac.in

Nazar Hussain & Atul Kumar (Eds.)

Fig. (1). Emil fischer.

Glycals are a class of carbohydrate derivatives that play a significant role in organic and medicinal chemistry. The term "glycals" is derived from combining "glycosides" and "alcohol", reflecting the structural feature of these compounds. Glycals are essential carbohydrates that contain a C=C double bond between C-1 and C-2 of the carbohydrate scaffold [3]. Glycals are versatile intermediates in the synthesis of complex carbohydrates and have found applications in medicinal chemistry, natural product synthesis, and many other areas of chemical science. They found their versatility and applicability in carbohydrate chemistry because of their stable transformations as well as unsaturation present at C-1 and C-2 positions [4]. They are unsaturated sugars that contain a double bond at the anomeric position, as shown in Fig. (**2**). The IUPAC name of the most common glycal is 1, 2-dideoxy-hex-1-eno-pyranose.

'Glycal' is a general term for all 1,2-unsaturated sugars, while the glycals of specific sugars have their own name from which they are derived *e.g.*, glucal from glucose, galactal from galactose, xylal from xylose, and so on (Fig. **3**).

Depending on the electronic nature and size of the substituents, as well as the presence of the ring oxygen atom, glycals can adopt either $^5H_4$ or $^4H_5$ conformation, as shown in Fig. (**4**).

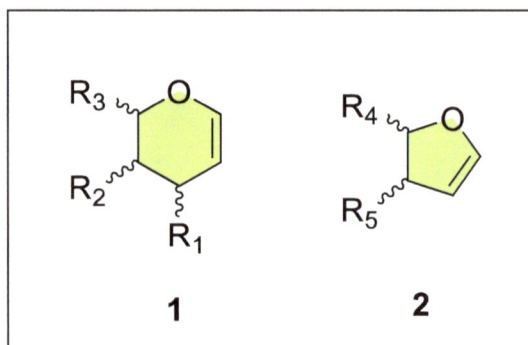

Fig. (2). General structure of pyranose and furanose glycals.

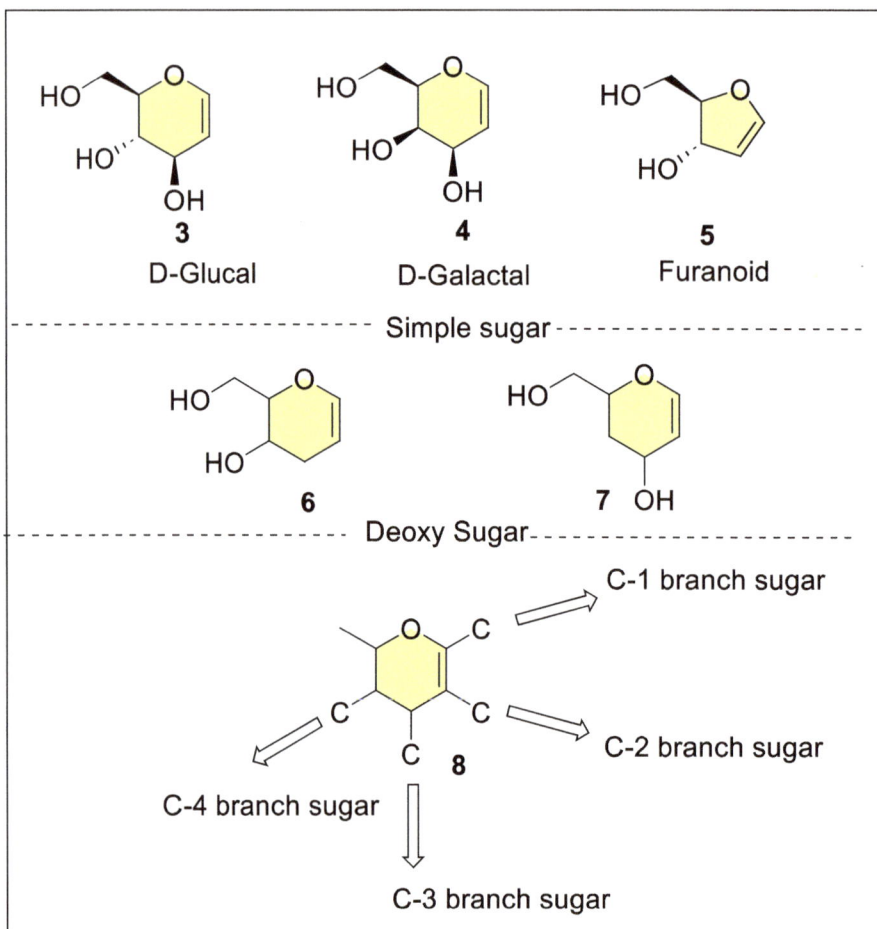

Fig. (3). Examples of some glycals.

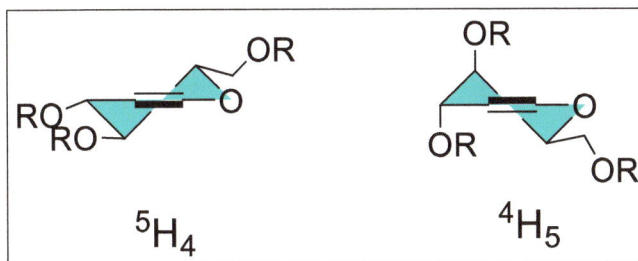

Fig. (4). Stable conformations of glycals.

Glycals can play an important role as a chiral starting material in various reactions like halogenation, epoxidation, ozonolysis, cycloaddition, transmetalation, hydrogenation, *etc.*, and also in the synthesis of different kinds of sugar derivatives like *O*-glycosides, *C*-glycosides, *S*-glycosides and *N*-glycosides and biologically relevant natural products by employing glycals [5]. During the transformation of glycals into other useful molecules, the hydroxyl groups at the 3, 4, and 6 positions of glycals may not be stable in many reaction conditions. So, they need to be protected before being subjected to various reactions. There are several hydroxyl protection strategies available in the literature *e.g.*, ether and ester protections, which are very handy and widely used in glycal reactions. Ether protection strategies include benzyl, ethyl, methyl, propyl, butyl, allylic protections, *etc.*, while ester protection strategies include acetyl, benzoyl, *etc.* Some examples of ether and ester-protected glycals are shown in Fig. (5).

PREPARATION OF SIMPLE GLYCALS

The first effective synthesis of 3, 4, 6-tri-*O*-acetyl-D-glucal **31** was reported by Emil Fischer and Zach from D-glucose **29** *via* 2, 3, 4, 6-tetra-*O*-acetyl glucopyranosyl bromide **30**, as shown in (Scheme **1**) [6]. It has been a traditional method for the synthesis of glycals on a bulk scale since 1913. The drawback of this methodology is that it fails to synthesize furanoid glycals under similar reaction conditions.

Carbohydrate chemists across the globe have devised several methods for the preparation of glycals from time to time. However, some of the most widely used methods like reductive elimination of phenyl thioglycosides [7], Danishefsky's hetero Diels-Alder approach [8], synthesis from glycosyl sulfoxides [9] and glycosyl sulfones [10], electrochemical methods [11], ring-closing metathesis [12], *etc.*, are described in the following section below.

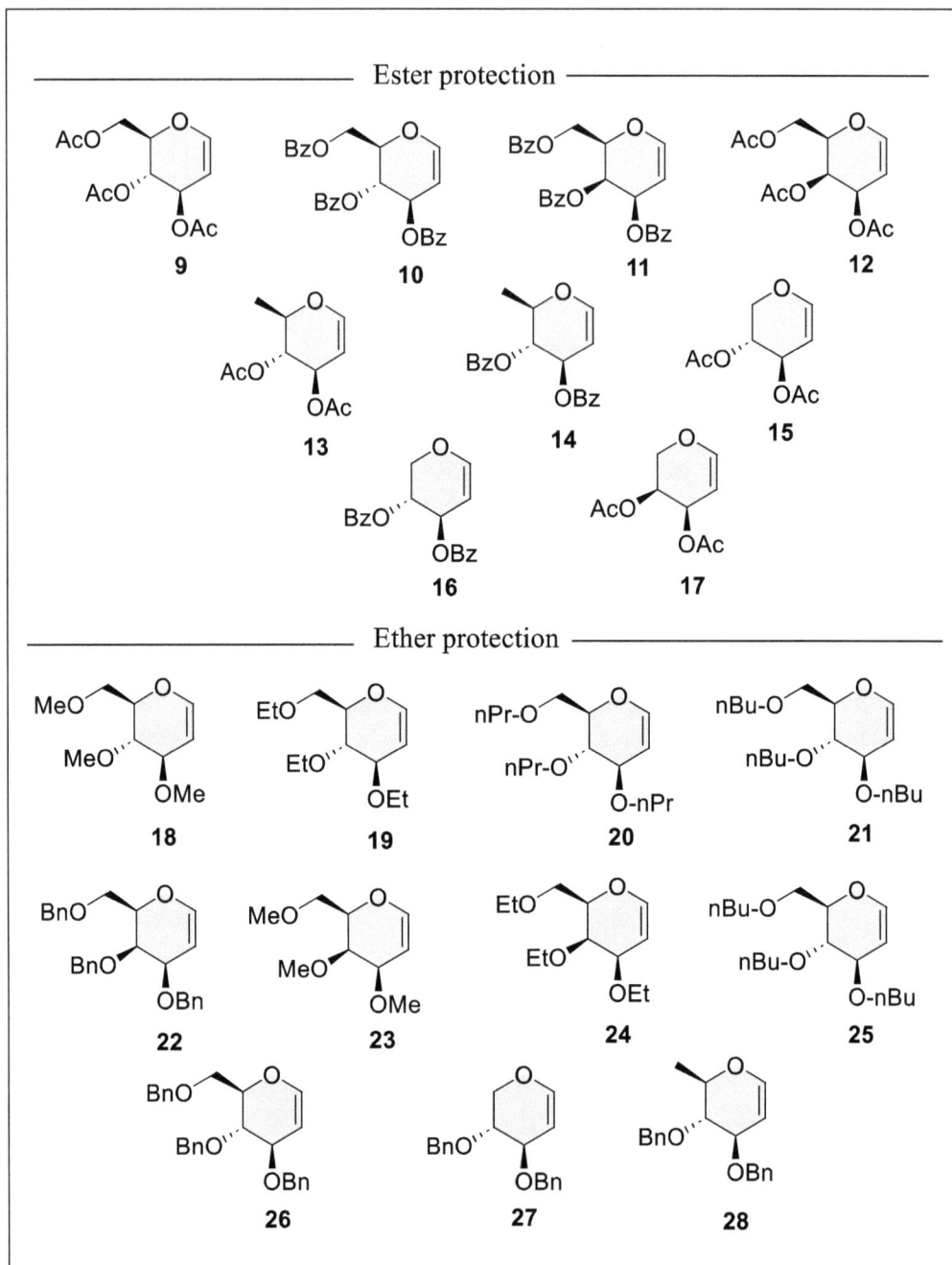

Fig. (5). Glycals protected with different ether and ester protecting groups.

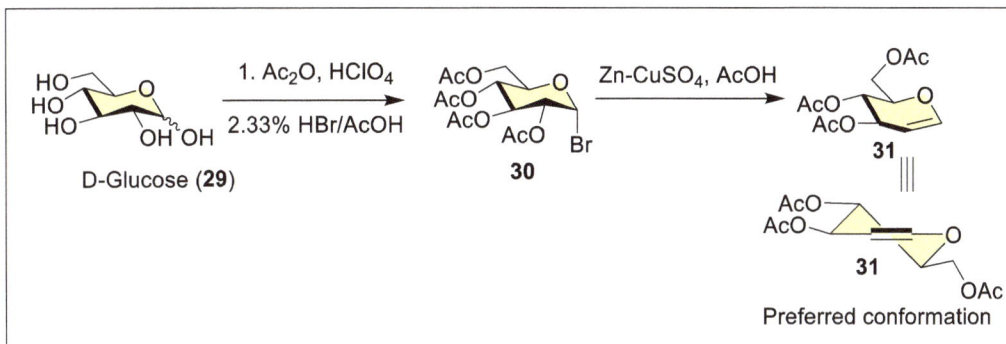

Scheme 1. Fischer-Zach method for the synthesis of D-glucal.

SYNTHESIS OF GLYCALS FROM GLYCOSYL SULFOXIDES

Preparation of Pyranoid Glycals

Firstly, galactose-derived sulfoxide **34** is synthesized from tetraacetate **32** in three steps: (i) Zemplen deacetylation, (ii) methylation, and (iii) oxidation of sulfide **33** with *m*-CPBA at a low temperature. Galactosylsulfoxide **34** is then treated with 3.0 equiv. of *n*-BuLi in THF at -78 °C for 20 minutes, which results in a clean reaction mixture from which protected galactal **35** is isolated in 59% yield, as illustrated in (Scheme **2**) [13 - 16].

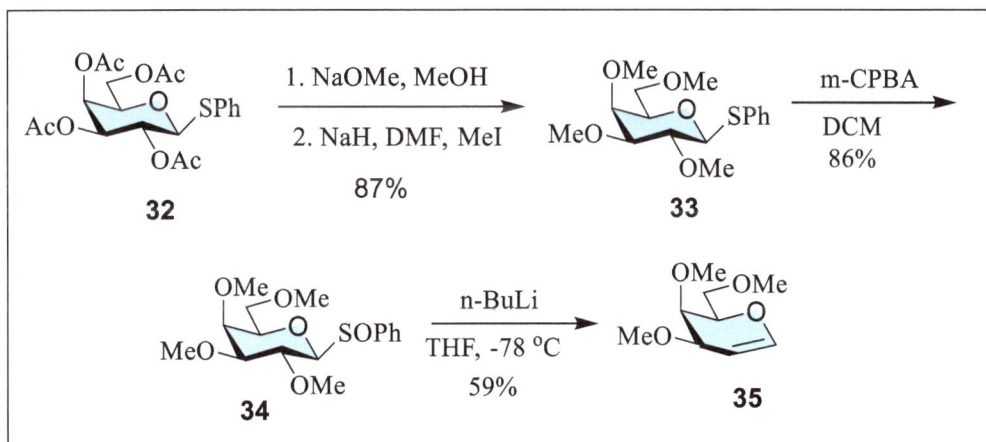

Scheme 2. Preparation of galactal sulfoxide.

Following the similar reaction sequence, glycosyl sulfoxides **39**, **40,** and **41,** exhibiting various stereochemical variations, are prepared from thioglysosides of D-galactose, D-mannose, and D-glucose **36**, **37,** and **38,** respectively (Scheme **3**).

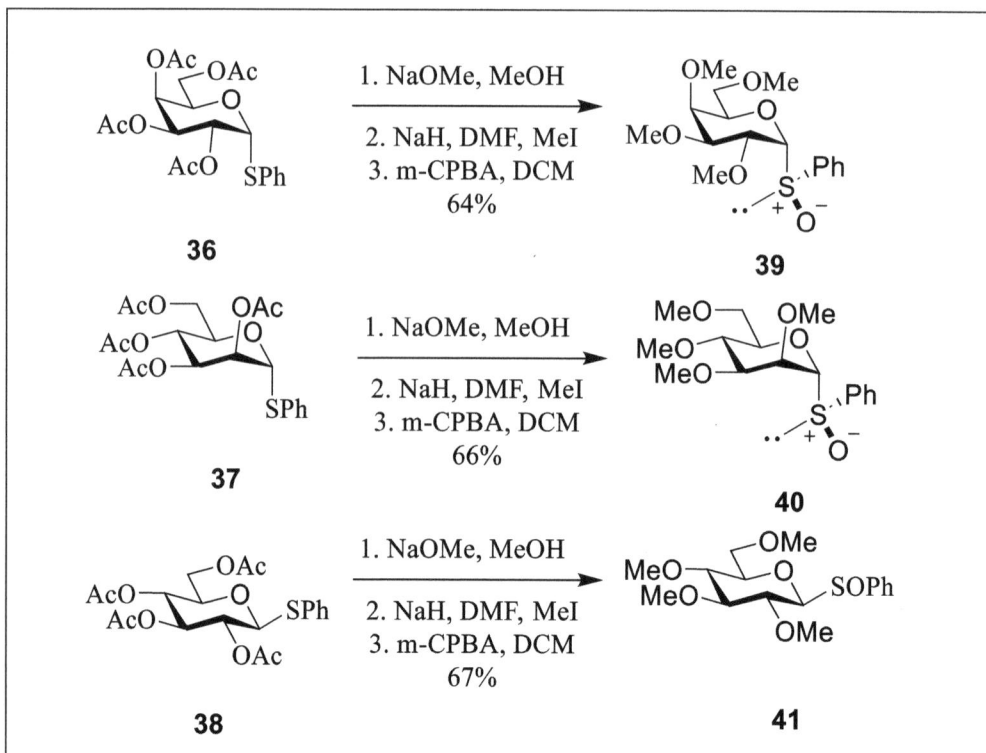

Scheme 3. Preparation of glycosyl sulfoxides.

Finally, these glycosyl sulfoxides **39-41** are treated with *n*-BuLi (2.0 equiv.) in a THF solution at a low temperature to generate the desired glycals **35** and **42**, as shown in (Scheme 4). Both the glycosyl sulfoxides **40** and **41** produce the same glycal as the C-2 substituent whose orientation distinguishes them from the final product *i.e.*, glycal **42**.

Preparation of Furanoid Glycals

The methodology described above for the preparation of pyranoid glycals is further extended successfully for the preparation of furanoid glycals. Furanosyl sulfoxide **44** is initially synthesized from hemiacetal **43** by thioglycosidation, followed by controlled oxidation. The treatment of sulfoxide **44** with *n*-BuLi generates the desired furanoid glycal **45** in 79% yield, as depicted in (Scheme 5).

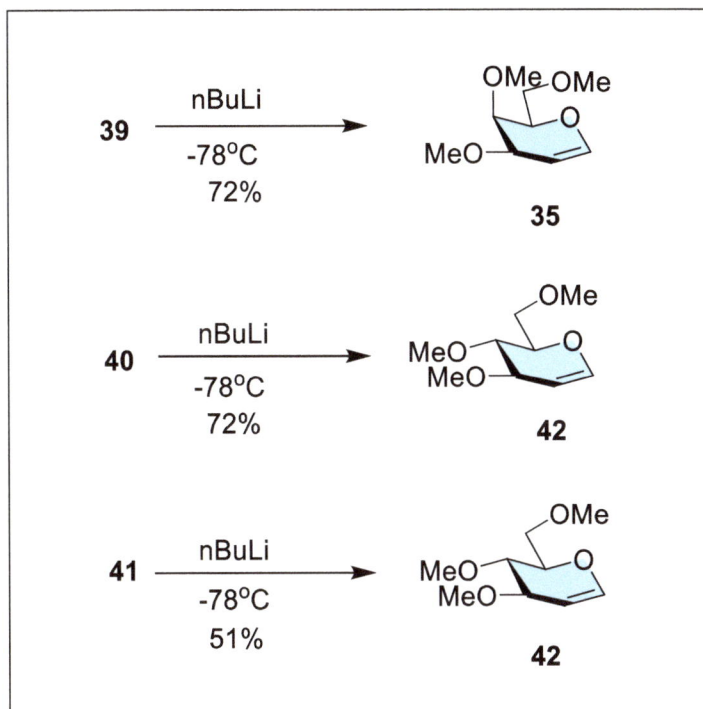

Scheme 4. Formation of glycals using *n*-BuLi in THF.

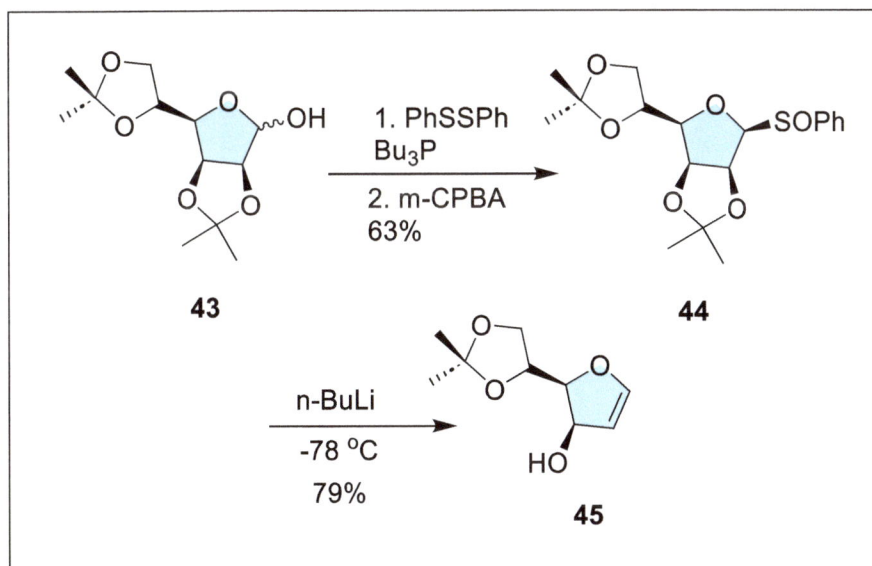

Scheme 5. Formation of furanoid glycals.

SYNTHESIS OF GLYCALS FROM THIOGLYCOSIDES OR GLYCOSYL SULFONES

In this methodology, thioglycosides and glycosyl sulfones (**46-50**) shown below are used for the preparation of glycals (Fig. **6**).

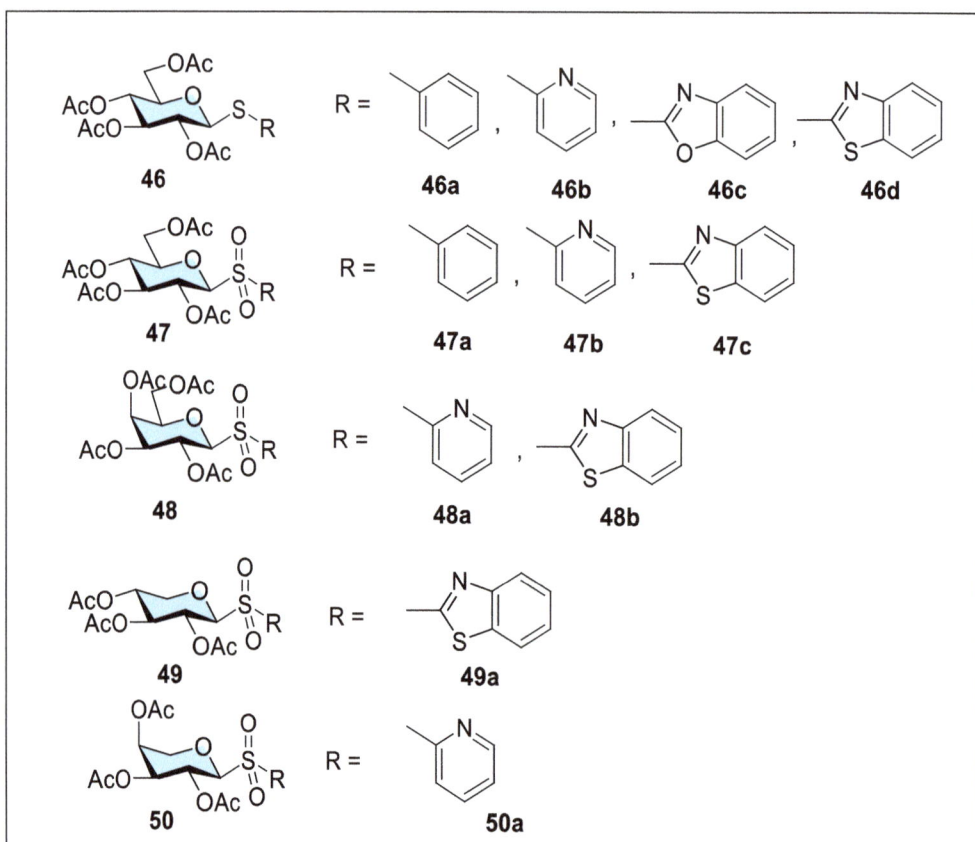

Fig. (6). Thioglycosides and glycosyl sulfones.

In this method, thioglycosides and glycosyl sulfones mentioned above are treated with reactive chromium (II) complex, [Cr (EDTA)]$^{-2}$. Successful synthesis of a variety of glycals (**52**) is achieved using this reaction. The reaction proceeds through a glycosyl-chromium (II) intermediate complex, as described schematically in (Scheme **6**).

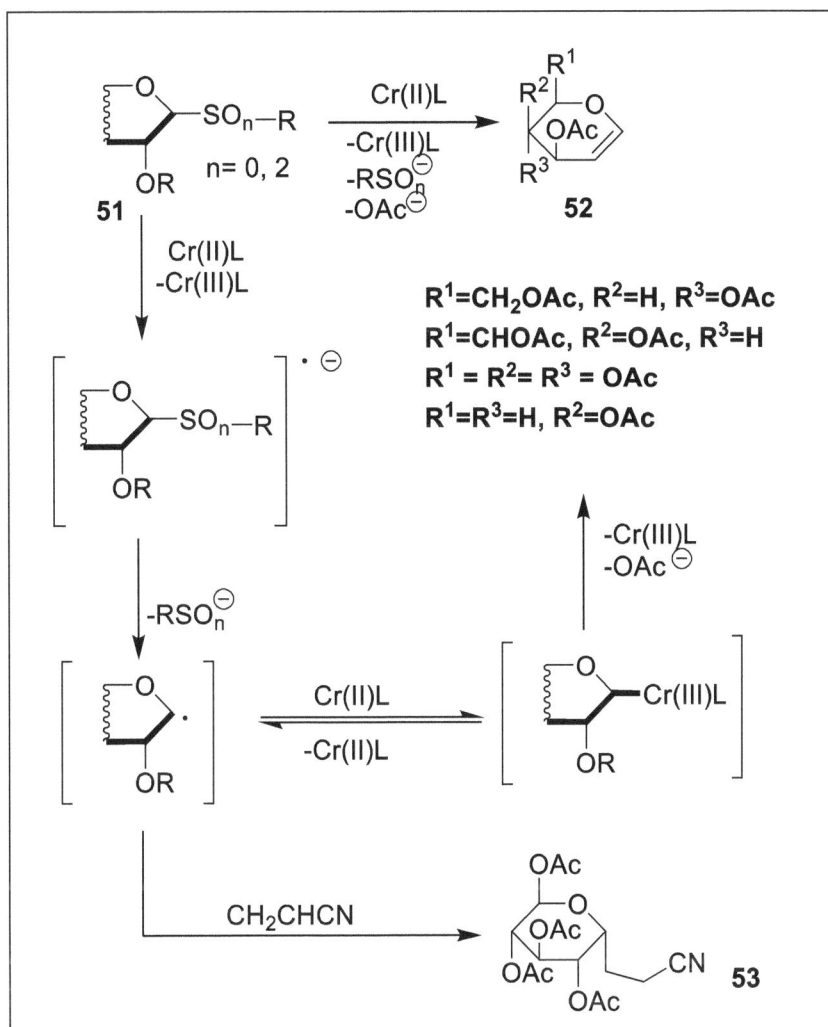

Scheme 6. Synthesis of glycals synthesis using chromium (II) complex.

SYNTHESIS OF GLYCALS BY ELECTROCHEMICAL METHOD

Traditionally, sugars are synthesized by the electrochemical reduction of glycosyl halide in the presence of various metallic or organometallic reagents using a mercury cathode [17]. Later on, R. Daniel's group reported [11] the synthesis of glycals by utilizing controlled current electrolysis. They performed their investigation by using reticulated vitreous carbon (RVC) cathodes and a consumable zinc anode and successfully achieved the rapid conversion of **54** into **55** (Scheme **7**). Various glycosyl halides were subjected to this electrochemical reduction, and the corresponding glycals were prepared, as shown in (Scheme **8**).

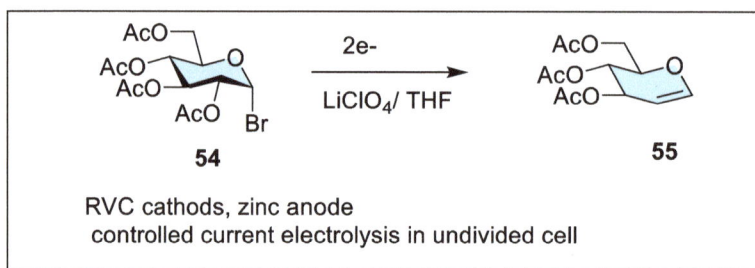

Scheme 7. Synthesis of glycals using an electrochemical method.

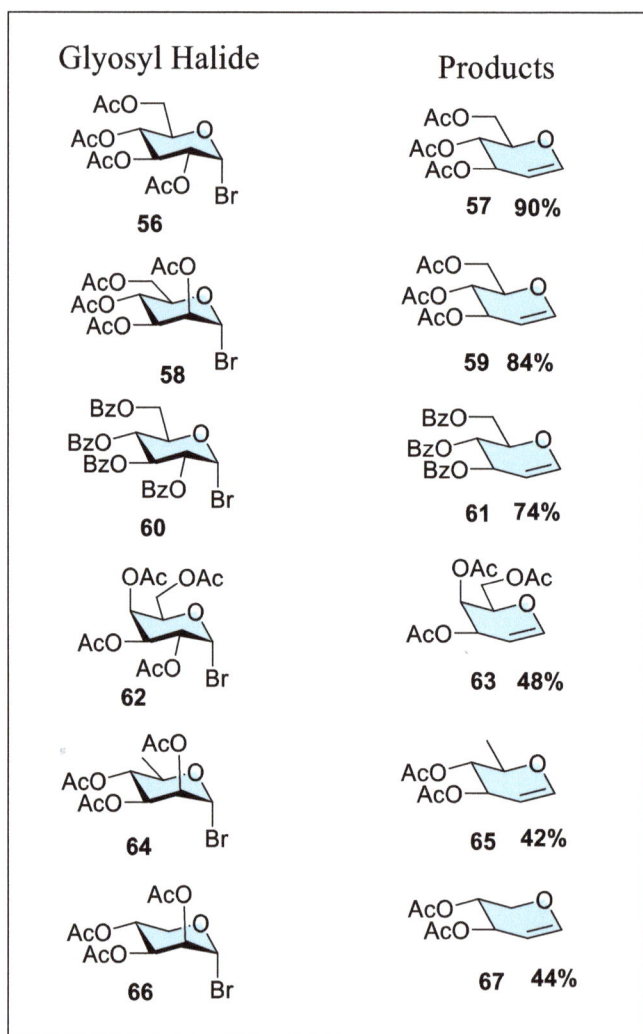

Scheme 8. Glycosyl halides and glycals.

SYNTHESIS OF GLYCALS USING TITANIUM(III) REAGENT

There are several methods for the synthesis of glycals from glycosyl halides [18, 19]. However, the most important method involves the preparation of glycals using Ti(II)-complex. In this method, a Ti(II) complex **68** (Fig. **7**) readily dehalogenates alkyl halides *via* selective halogen abstraction [20]. This reagent is very interesting in the selective transformation of glycosyl halides. The reaction between peracetylated glycosyl bromide and **68** forms corresponding glycals in very good yields [20 - 22]. One interesting fact is that elimination occurs regardless of the stereochemistry at C-2 *i.e.*, glucosyl and mannosyl bromide are converted into glucal in very good yields. The reaction follows a radical-based mechanism where initial halide abstraction by Ti(II) produces a 1-glycosyl radical. The second equivalent of Ti(II) traps this radical and gives (glycosyl) titanium (IV) species after β-elimination of $Cp_2Ti(OAc)$ produces the desired glucal product **72** (Scheme **9**). By using this methodology, various types of pyranoid glycals are synthesized, but in the case of furanoid glycals, there are difficulties in isolating them under these reaction conditions. Usually, for the synthesis of glycals, it is essential to have a very good leaving group in the C-3 position.

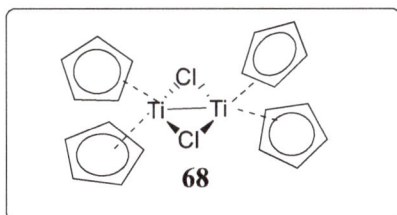

Fig. (7). Ti(II) complex.

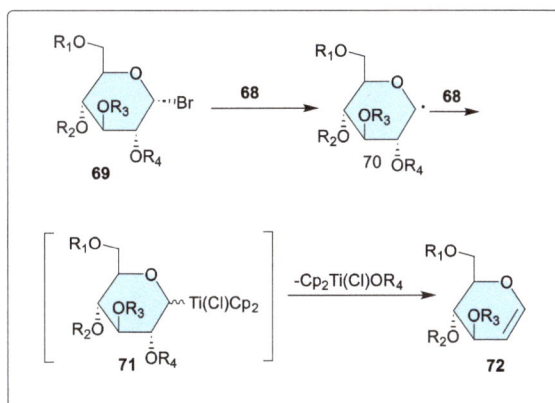

Scheme 9. Synthesis of glycals by using titanium complex.

TRANSFORMATIONS OF GLYCALS

As previously stated, glycals are useful chiral starting materials for many reactions. A brief summary of glycal transformations is given below (Scheme **10**).

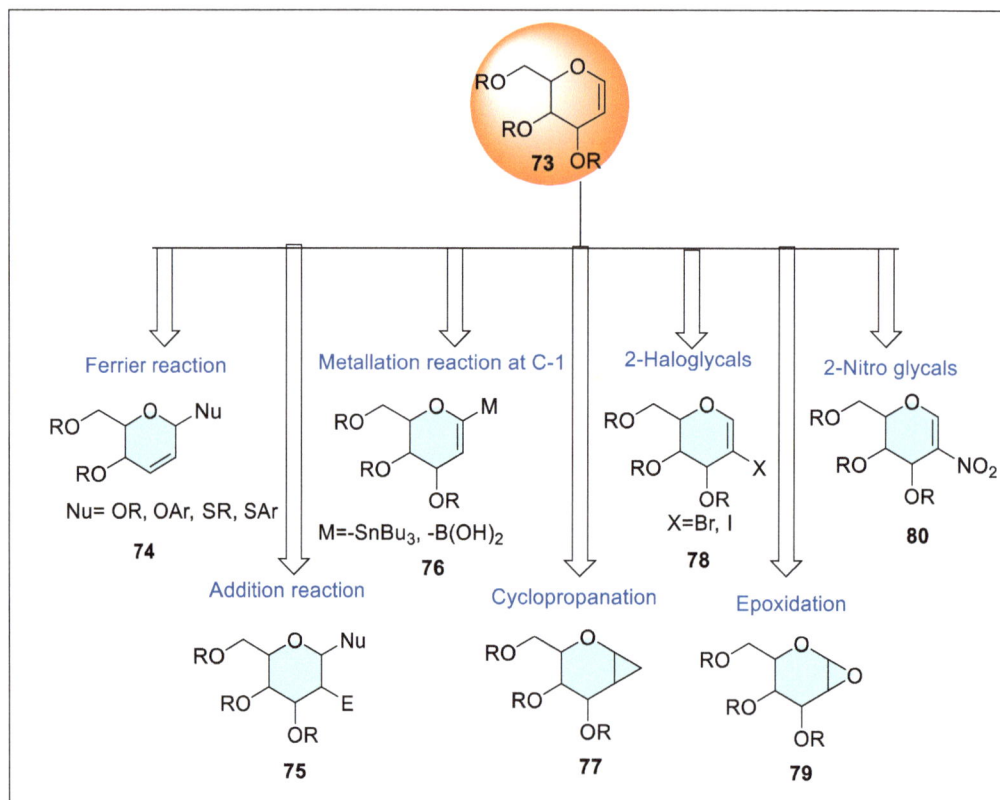

Scheme 10. Transformations of glycals.

The reactivity of glycals is due to the unsaturation present inside the ring. The reactions of glycals include rearrangement reactions, addition reactions, cycloadditions, epoxidations, transmetalations, *etc.*, as depicted in (Scheme **10**). A very famous reaction in carbohydrate chemistry is Ferrier rearrangement, discovered by Ferrier in the sixties, which involves the allylic rearrangement in the presence of Lewis acid to produce a rearranged product **74**. This transformation finds application in the synthesis of natural products or drug molecules.

GLYCALS IN SYNTHETIC AND NATURAL PRODUCTS

Synthetic and natural glycosides contain carbohydrate moiety, and glycals are the best easily available chiral starting units for the generation of the sugar part of these glycosides, thus playing a significant role in drug synthesis. *C*-glycosides, which possess various biological properties, caught the interest of medicinal chemists and researchers, leading to the development of several commercially available *C*-glycoside drugs, as shown in (Scheme **8**). These molecules are extensively used as anti-diabetic drugs *e.g.*, Dapagliflozin **91** [23], Canagliflozin **92** [24], and Empagliflozin **93** [25], and are used in the treatment of type-2 diabetes (Fig. **8**).

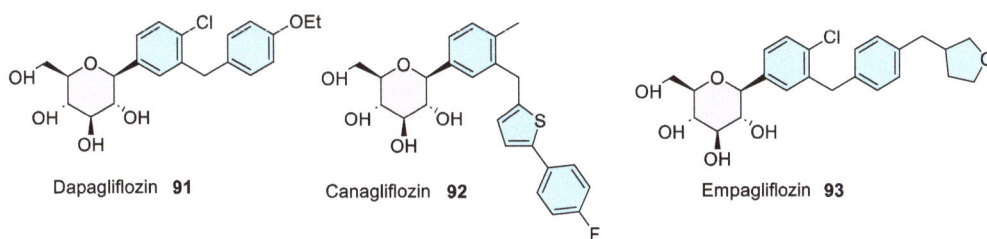

Fig. (8). Synthetic *C*-glycosides with noticeable biological activities.

Glycals are even more extensively used in the synthesis of a wide range of biologically significant natural product glycosides like C-Man-Trp **81**, Aloin **82**, Salmochelin SX **83**, Showdomycin **84**, Formycin **85**, (+)-Varitriol **86**, Vitexin **87**, Orientin **87**, Isovitexin **88**, Isoorintin **89**, Aquayamycin **89**, Cassialoin **90** *etc.*, as depicted in Fig. (**9**) [26].

EXO-GLYCALS

C-glycosyl compounds having an *exo*-cyclic carbon-carbon double bond at the anomeric center are known as *exo*-glycals and are sometimes also referred to as *C*-glycosylidene [27]. *Exo*-glycal could be of great interest as precursors of *C*-glycosides if the double bond is reduced with high stereo-control. The presence of an oxygen atom inside the ring strongly influences the reactivity of this double bond, so various properties should be expected from these olefins. The discovery of several direct and stepwise methods for the formation of a carbon-carbon double bond at the anomeric center of glycal has increased the availability of these unsaturated compounds and strengthened their role in organic synthesis. Various strategies used by carbohydrate chemists in the synthesis of *exo*-glycals **95-100** are summarized in (Scheme **11**).

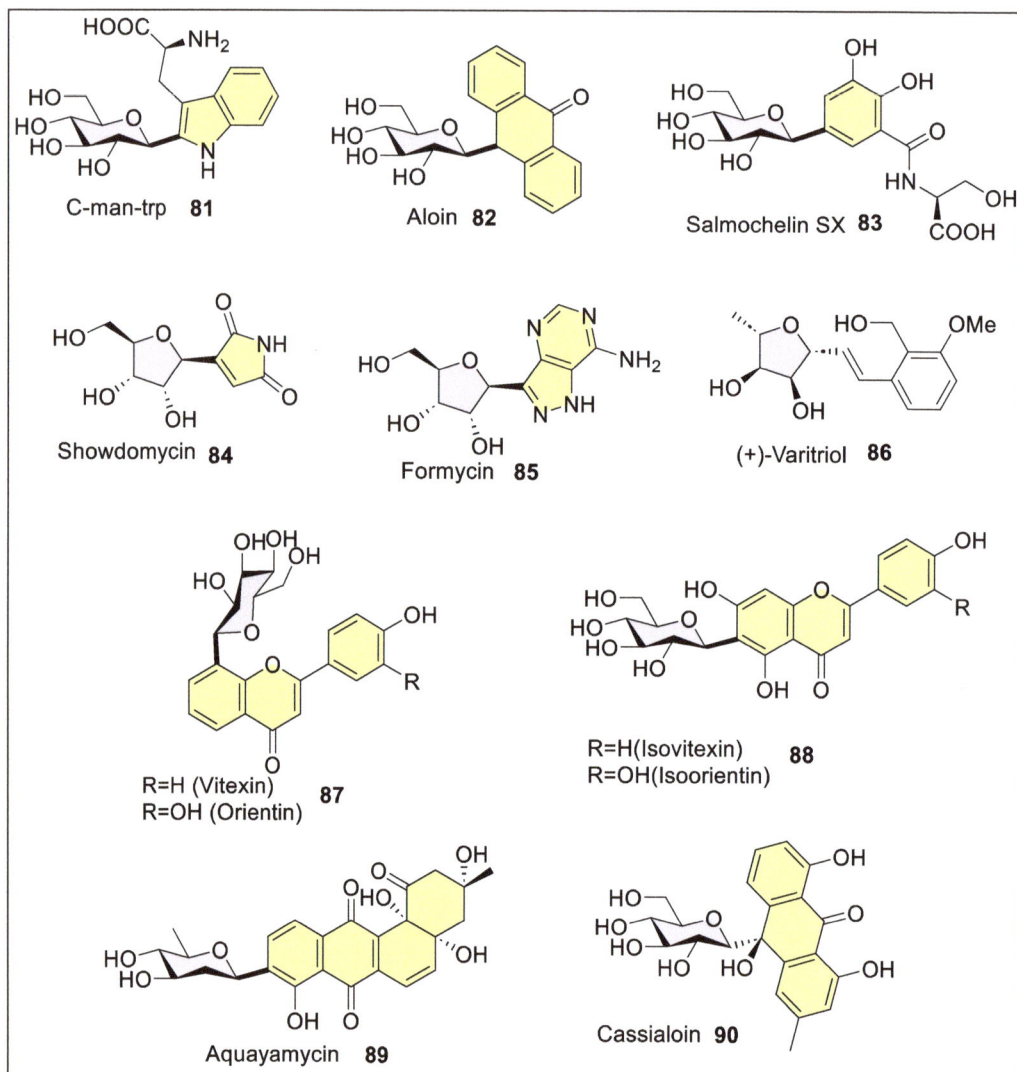

Fig. (9). Medicinally important natural products *C*-glycosides.

Synthesis of *Exo*-glycals from Sugar Lactones

The first *exo*-glycal was synthesized in 1975 by Bischofberger *et al.*, who studied the reaction of sugar lactones with ethyl isocyanoacetate [28]. After that, several new and more efficient methods were reported. The first example of *C*-methylene derivative was reported by Brockhaus and Lehmann in 1977 [29]. This preparation was based on a stepwise procedure in which the *C*-glycosyl derivative attached to an iodine atom with subsequent elimination according to the well-established reaction condition, which led to the formation of unsaturated sugar.

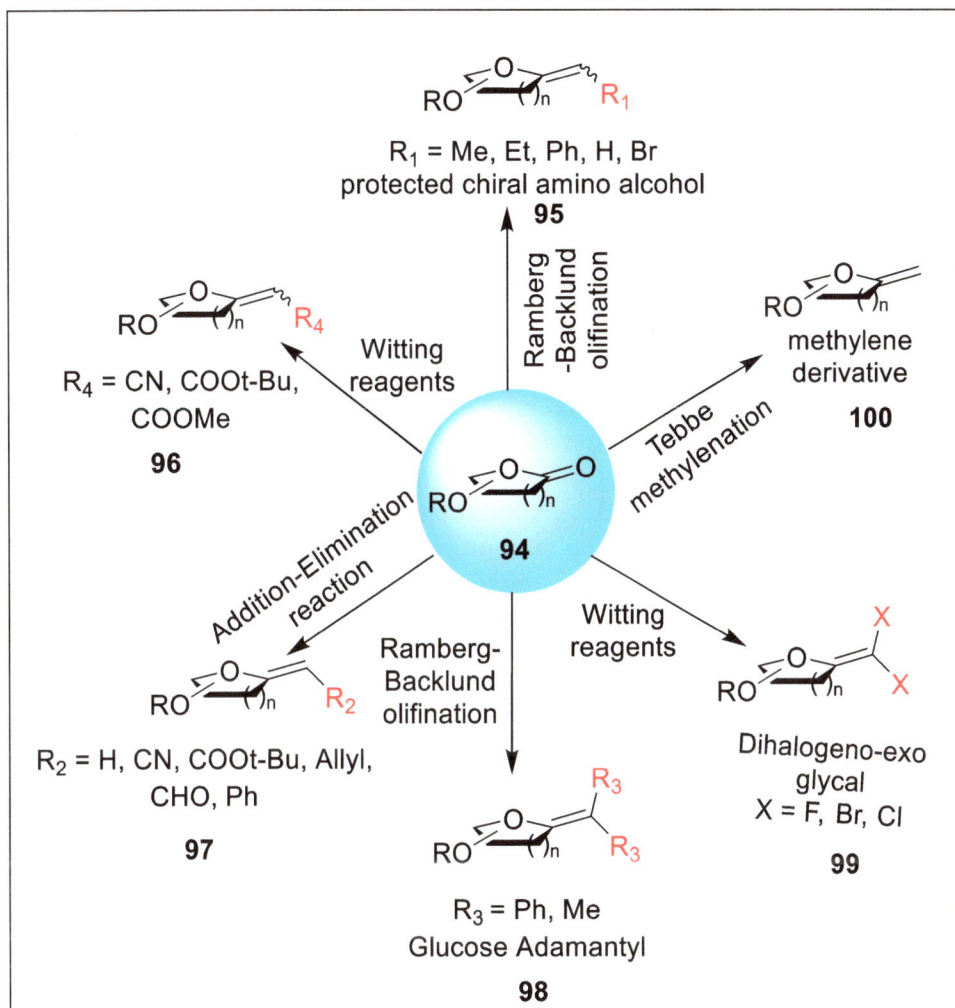

Scheme 11. Strategies reviewed for the synthesis of substituted *exo*-glycals.

As already mentioned, the reaction of ethyl isocyanoacetate with sugar lactone **101** as a starting compound under the conditions depicted in (Scheme **12**) produced the first *exo*-glycal **102**. Further catalytic hydrogenation of *exo*-glycal **102** converted it into sugar amino acid **103**, as shown in (Scheme **12**). However, this research work could not be explored further due to the formation of acyclic sugar oxazole in the presence of potassium ethyl isocyanate salt.

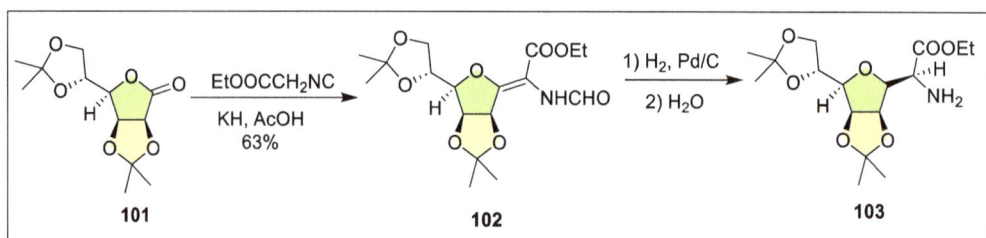

Scheme 12. Synthesis of *exo*-glycal using lactone and ethyl isocyanoacetate.

The first example of a modified Julia olefination in order to prepare methylene *exo*-glycal was reported by Gueyrard *et al.* [30] in 2005 by using sugar-derived lactone (Scheme **13A**). In the presence of a strong non-nucleophilic base like LiHMDS, a reaction occurs between tri-*O*-benzyl-D-arabinolactone **104** and methyl benzothiazoylsulfone **105** to form an expected intermediate α-heteroarylsulfone hemiketal **106**. In the presence of DBU, Smile rearrangement occurs, which gives the *exo*-glycal or anomeric methylene derivative **107**.

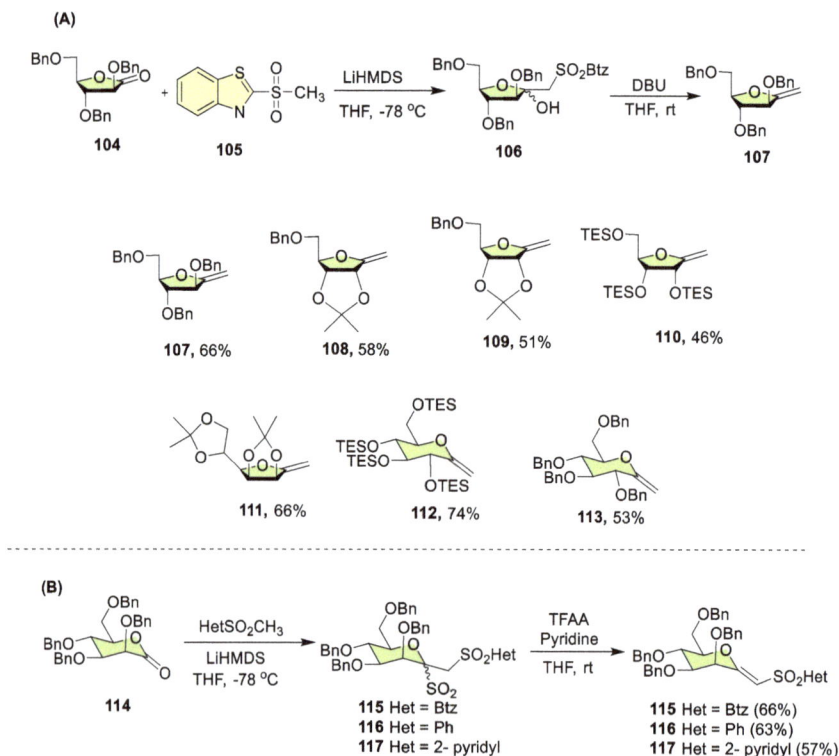

Scheme 13. Synthesis of *exo*-glycals from sugar lactones by Julia olefination.

Various *exo*-glycals **107-113** are synthesized in moderate to good yields by using these reaction conditions. These conditions are tolerated by many protecting groups (benzyl, silyl, and isopropylidene) and different sugars (D-arabino, D-manno, D-ribo, D-gluco, and D-erythro). In place of DBU, trifluoroacetic anhydride and pyridine can be used to synthesize *C*-glucosyl vinyl sulfones **115-117**. This method was reported by Gueyrard's group [31] to provide exclusive *Z*-selectivity in the product, as depicted in (Scheme **13B**).

Recently, Kaszás *et al.* [32] have reported Pd-catalyzed coupling of *O*-peracylated 2,6-anhydro-aldose tosylhydrazones **118** with aryl bromide to synthesize aryl-substituted *exo*-glycals **119-128**. According to the nature of protecting groups, the reactions gave a mixture of diastereomers with good yields, as described in (Scheme **14**). The sugar configuration did not seem to influence the outcome of the transformations, so such methods provide a new methodology for the preparation of aryl-substituted *exo*-glycals.

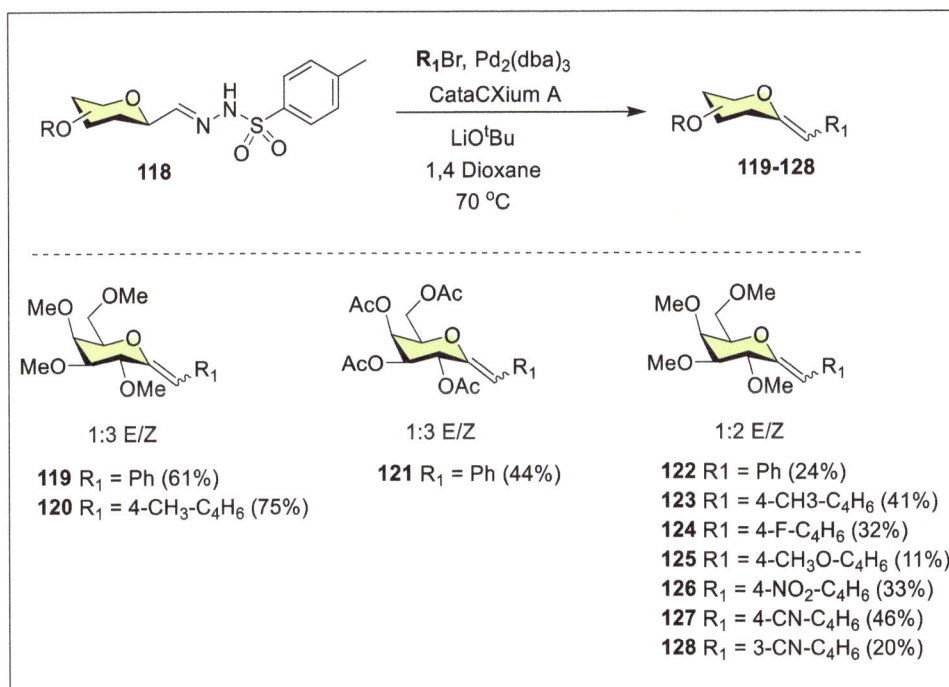

1:3 E/Z

119 R_1 = Ph (61%)
120 R_1 = 4-CH$_3$-C$_4$H$_6$ (75%)

1:3 E/Z

121 R_1 = Ph (44%)

1:2 E/Z

122 R1 = Ph (24%)
123 R1 = 4-CH3-C$_4$H$_6$ (41%)
124 R1 = 4-F-C$_4$H$_6$ (32%)
125 R1 = 4-CH$_3$O-C$_4$H$_6$ (11%)
126 R_1 = 4-NO$_2$-C$_4$H$_6$ (33%)
127 R_1 = 4-CN-C$_4$H$_6$ (46%)
128 R_1 = 3-CN-C$_4$H$_6$ (20%)

Scheme 14. Synthesis of *exo*-glycals using Pd-catalyzed coupling.

Synthesis and Chemoselective Oxidation of Thiomethylphosphono-*exo*-glycals

Frédéric *et al.* reported the synthesis of sulfonylated phosphono-*exo*-glycals [33]. A lithiated phosphonate group having an electron-donating nature was added to various lactones because the addition of a lithiated phosphonates group bearing a sulfonyl directly is impossible. Subsequent dehydration occurs in the next step to provide *exo*-glycals **130-133,** as demonstrated in (Scheme **15**). The isomeric product is obtained in good yields, and *Z/E* ratios ranging from 60:40 to 96:4 are achieved. After the synthesis of sulfonylated phosphono-*exo*-glycal, its oxidation is performed chemo selectively using two equivalents of *m*-chloroperoxybenzoic acid (*m*-CPBA) in dichloromethane solvent at 0 °C, and the desired compounds **134-138** are formed.

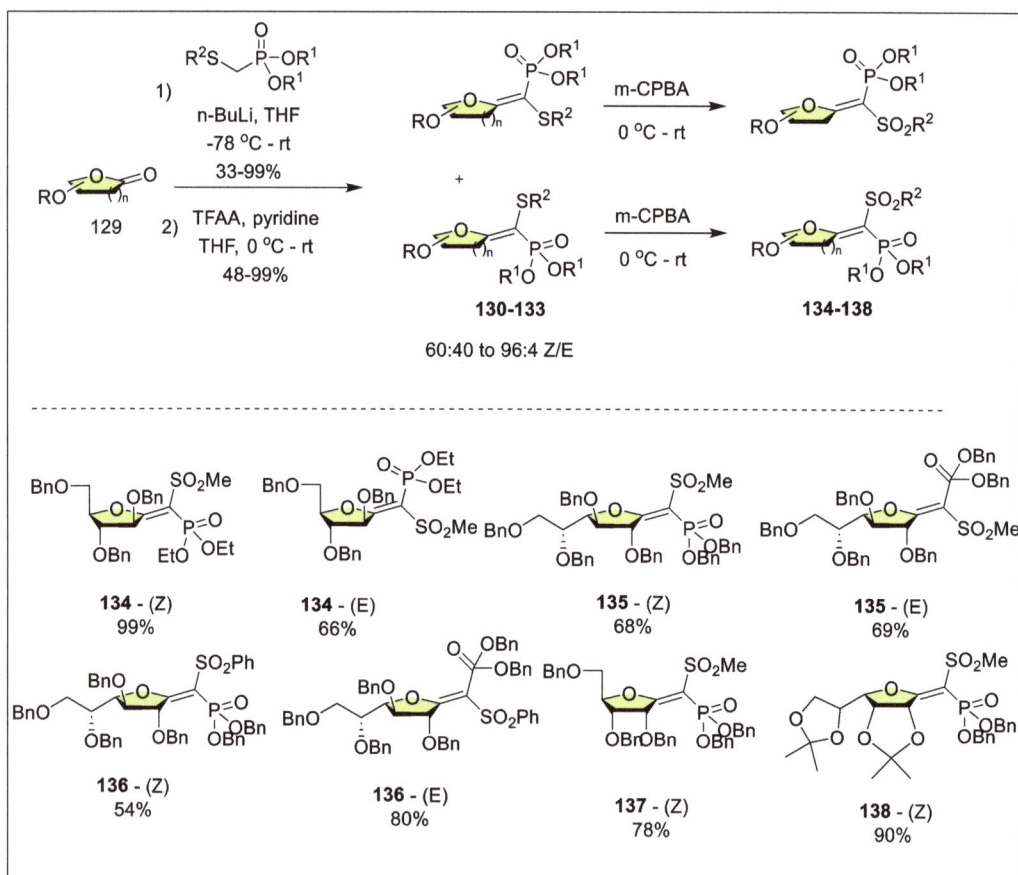

Scheme 15. Synthesis and oxidation of thiomethylphosphono-*exo*-glycals.

Synthesis of Difluoro *Exo*-glycals

In 2015, Gueyrard's group [34] reported the synthesis of difluoro *exo*-glycals **142** using difluoromethyl-2-pyridyl sulfone **140** and lactones **139,** as shown in (Scheme **16A**). Julia modified-reagent **140** and lactones are coupled to each other in the presence of LiHMDS and BF$_3$·OEt$_2$ in THF at low temperatures, and a hemiketal intermediate **141** is produced. The rearrangement mediated by DBU does not take place in **141,** so the desired *exo*-glycals are synthesized by microwave irradiation at 140 °C for 1 h in toluene in the absence of a base (Scheme **16A**). This method provides good efficiency on various lactones except for silyl-protected substrates. Chen *et al.* [35] also synthesized difluoro *exo*-glycal **142** through a two-step reaction sequence where basic conditions are used in the second step (Scheme **16B**).

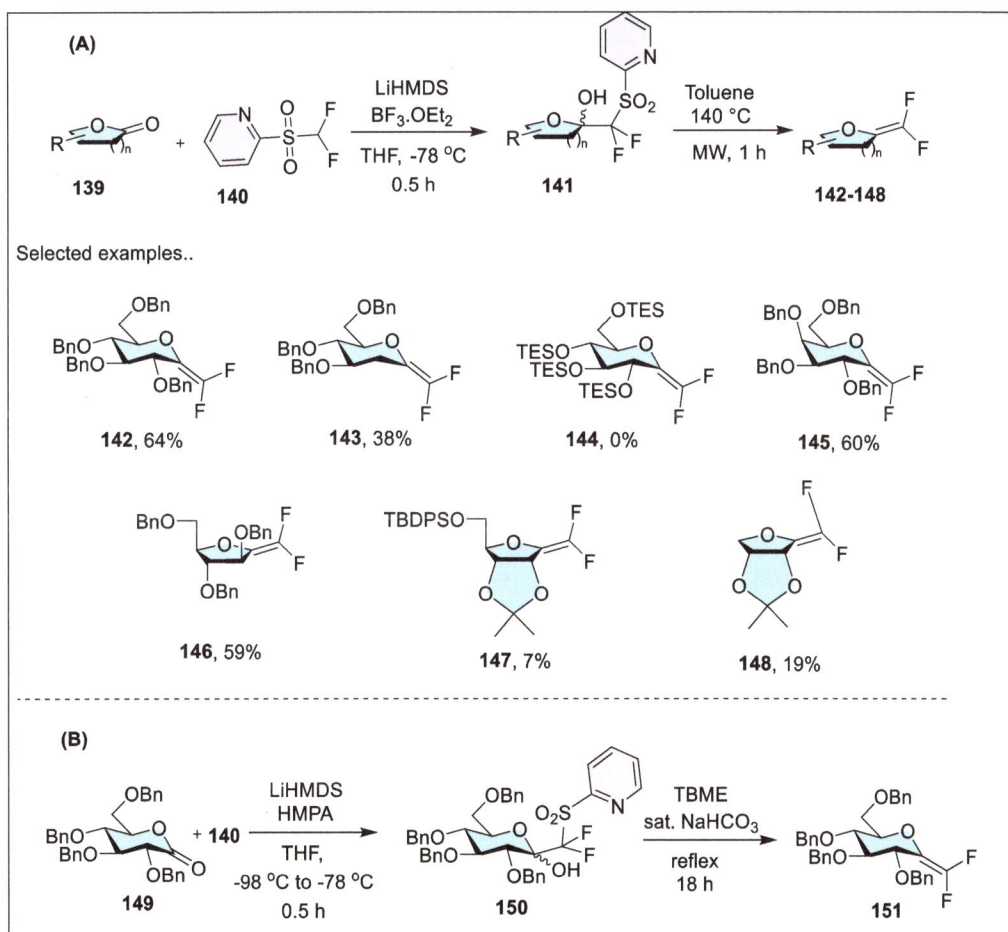

Scheme 16. Synthesis of difluoro *exo*-glycals.

CONCLUSION

In conclusion, we delved into the fascinating realm of unsaturated sugars, exploring both *endo*- and *exo*-glycals. These compounds serve as pivotal building blocks in the synthesis of various valuable products. We navigated through their reactivity and diverse transformations to achieve compounds of significant utility. Moreover, we meticulously outlined various strategies employed for the synthesis of these crucial sugar derivatives, offering readers a comprehensive understanding of their synthetic routes. Furthermore, we shed light on the conversion of these sugar derivatives into glycosides, providing a glimpse into their potential applications and importance in organic chemistry. Ultimately, this chapter serves as a cornerstone by offering a panoramic view that lays the groundwork for a deeper comprehension of subsequent chapters within the book.

REFERENCES

[1]   Tian, Q.; Xu, L.; Ma, X.; Zou, W.; Shao, H. Stereoselective synthesis of 2-C-acetonyl-2-deoxy- D-galactosides using 1,2-cyclopropaneacetylated sugar as novel glycosyl donor. *Org. Lett.,* **2010**, *12*(3), 540-543.
        [http://dx.doi.org/10.1021/ol902732w] [PMID: 20041707]

[2]   Levy, D. E., Fügedi, P., *The Organic Chemistry of Sugars*; Eds.; CRC Press, **2005**.
        [http://dx.doi.org/10.1201/9781420027952]

[3]   Ernst, B.; Hart, G.W.; Sinaý, P. *Carbohydrates in Chemistry and Biology*; Wiley, **2000**.
        [http://dx.doi.org/10.1002/9783527618255]

[4]   Ferrier, R.J. *Adv. Carbohydr. Chem. Biochem.,* **1969**.

[5]   Zhao, J.; Wei, S.; Ma, X.; Shao, H. A simple and convenient method for the synthesis of pyranoid glycals. *Carbohydr. Res.,* **2010**, *345*(1), 168-171.
        [http://dx.doi.org/10.1016/j.carres.2009.10.003] [PMID: 19892323]

[6]   Roth, W.; Pigman, W. *Methods in Carbohydrate Chemistry*; **1963**; *2*.

[7]   Boutureira, O.; Rodríguez, M.A.; Matheu, M.I.; Díaz, Y.; Castillón, S. General method for synthesizing pyranoid glycals. A new route to allal and gulal derivatives. *Org. Lett.,* **2006**, *8*(4), 673-675.
        [http://dx.doi.org/10.1021/ol052866l] [PMID: 16468739]

[8]   Danishefsky, S.J.; Bilodeau, M.T. Glycals in organic synthesis: the evolution of comprehensive strategies for the assembly of oligosaccharides and glycoconjugates of biological consequence. *Angew. Chem. Int. Ed. Engl.,* **1996**, *35*(13-14), 1380-1419.
        [http://dx.doi.org/10.1002/anie.199613801]

[9]   Gómez, A.M.; Casillas, M.; Barrio, A.; Gawel, A.; López, J.C. Synthesis of pyranoid and furanoid glycals from glycosyl sulfoxides by treatment with organolithium reagents. *Eur. J. Org. Chem.,* **2008**, *2008*(23), 3933-3942.
        [http://dx.doi.org/10.1002/ejoc.200800318]

[10]   Micskei, K.; Juhász, Z.; Ratković, Z.R.; Somsák, L. Reactivity of per-O-acetylated 1-thioglycosides and glycosyl sulfones towards chromium(II) complexes in aqueous medium. *Tetrahedron Lett.,* **2006**, *47*(34), 6117-6120.
        [http://dx.doi.org/10.1016/j.tetlet.2006.06.075]

[11]   Parrish, J.D.; Little, R.D. Electrochemical formation of glycals in THF. *Tetrahedron Lett.,* **2001**, *42*(42), 7371-7374.

[http://dx.doi.org/10.1016/S0040-4039(01)01544-1]

[12] Postema, M.H.D.; Calimente, D. Convergent preparation of 1,6-linked C-disaccharides *via* olefin metathesis. *Tetrahedron Lett.,* **1999**, *40*(26), 4755-4759.
[http://dx.doi.org/10.1016/S0040-4039(99)00815-1]

[13] Gildersleeve, J.; Pascal, R.A.; Kahne, D. Sulfenate intermediates in the sulfoxide glycosylation reaction. *J. Am. Chem. Soc.,* **1998**, *120*(24), 5961-5969.
[http://dx.doi.org/10.1021/ja980827h]

[14] Ferrier, R.J.; Furneaux, R.H. Synthesis of 1,2-trans-related 1-thioglycoside esters. *Carbohydr. Res.,* **1976**, *52*(1), 63-68.
[http://dx.doi.org/10.1016/S0008-6215(00)85946-7]

[15] Zemplén, G.; Kunz, A. Über die natriumverbindungen der glucose und die verseifung der acylierten zucker. *Berichte der Dtsch. Chem. Gesellschaft A B Ser,* **1923**, *56*(7), 1705-1710.

[16] Klemer, A.; Rodemeyer, G. Eine einfache Synthese von Methyl-4,6- *O* -benzyliden-2-desoxy-α-D-*erythro* -hexopyranosid-3-ulose. *Chem. Ber.,* **1974**, *107*(8), 2612-2614.
[http://dx.doi.org/10.1002/cber.19741070821]

[17] Spencer, R.P.; Schwartz, J. Titanium(III) reagents in carbohydrate chemistry: glycal and c-glycoside synthesis. *Tetrahedron,* **2000**, *56*(15), 2103-2112.
[http://dx.doi.org/10.1016/S0040-4020(99)01088-1]

[18] Roth, W.; Pigman, W. No Title. *Acad. Press New York,* **1963**, *2*, 405-408.

[19] Casillas, M. GoÂmez, A. M.; LoÂpez, J. C.; Valverde, S. No Title. *Synlett,* **1996**, 628-630.
[http://dx.doi.org/10.1055/s-1996-5553]

[20] Cavallaro, C.L.; Schwartz, J. A rapid synthesis of pyranoid glycals from glycosyl bromides. *J. Org. Chem.,* **1995**, *60*(21), 7055-7057.
[http://dx.doi.org/10.1021/jo00126a075]

[21] Spencer, R.P.; Cavallaro, C.L.; Schwartz, J. Rapid preparation of variously protected glycals using titanium(III). *J. Org. Chem.,* **1999**, *64*(11), 3987-3995.
[http://dx.doi.org/10.1021/jo982447k]

[22] Spencer, R.P.; Schwartz, J. Variously substituted glycals are readily prepared from glycosyl bromides using (Cp2TiCl)2. *Tetrahedron Lett.,* **1996**, *37*(25), 4357-4360.
[http://dx.doi.org/10.1016/0040-4039(96)00865-9]

[23] Zhu, F.; Rourke, M.J.; Yang, T.; Rodriguez, J.; Walczak, M.A. Highly stereospecific cross-coupling reactions of anomeric stannanes for the synthesis of *c*-aryl glycosides. *J. Am. Chem. Soc.,* **2016**, *138*(37), 12049-12052.
[http://dx.doi.org/10.1021/jacs.6b07891] [PMID: 27612008]

[24] Pałasz, A.; Cież, D.; Trzewik, B.; Miszczak, K.; Tynor, G.; Bazan, B. In the search of glycoside-based molecules as antidiabetic agents. *Top. Curr. Chem. (Cham),* **2019**, *377*(4), 19.
[http://dx.doi.org/10.1007/s41061-019-0243-6] [PMID: 31165274]

[25] Woo, V.C. Cardiovascular effects of sodium-glucose cotransporter-2 inhibitors in adults with type 2 diabetes. *Can. J. Diabetes,* **2020**, *44*(1), 61-67.
[http://dx.doi.org/10.1016/j.jcjd.2019.09.004] [PMID: 31839265]

[26] Xiao, M.; Wu, W.; Wei, L.; Jin, X.; Yao, X.; Xie, Z. Total synthesis of (−)-isatisine A *via* a biomimetic benzilic acid rearrangement. *Tetrahedron,* **2015**, *71*(22), 3705-3714.
[http://dx.doi.org/10.1016/j.tet.2014.09.028]

[27] Williams, M. Report of the international union of pure and applied chemistry: 1996. *Chem. Int. Newsmag. IUPAC,* **1997**, *19*(4).

[28] Hall, R.H.; Bischofberger, K.; Eitelman, S.J.; Jordaan, A. Synthesis of c-glycosyl compounds. part 1. reaction of ethyl isocyanoacetate with 2,3:5,6-di-o-isopropylidene-d-mannono-1,4-lactone. *J. Chem.*

Soc., Perkin Trans. 1, **1977**, (7), 743.
[http://dx.doi.org/10.1039/p19770000743]

[29]   Brockhaus, M.; Lehmann, J. The conversion of 2,6-anhydro-1-deoxy-d-galacto-hept-1-enitol into 1-deoxy-d-galacto-heptulose by β-d-galactosidase. *Carbohydr. Res.,* **1977**, *53*(1), 21-31.
       [http://dx.doi.org/10.1016/S0008-6215(00)85451-8] [PMID: 912679]

[30]   Gueyrard, D.; Haddoub, R.; Salem, A.; Bacar, N.S.; Goekjian, P.G. Synthesis of methylene *exo*glycals using a modified julia olefination. *Synlett,* **2005**, (3), 520-522.
       [http://dx.doi.org/10.1055/s-2005-862364]

[31]   Gueyrard, D.; Fontaine, P.; Goekjian, P. Synthesis of *c* -glucoside *endo* -glycals from *c* -glucosyl vinyl sulfones. *Synthesis,* **2006**, *2006*(9), 1499-1503.
       [http://dx.doi.org/10.1055/s-2006-926422]

[32]   Kaszás, T.; Ivanov, A.; Tóth, M.; Ehlers, P.; Langer, P.; Somsák, L. Pd-catalyzed coupling reactions of anhydro-aldose tosylhydrazones with aryl bromides to produce substituted *exo* -glycals. *Carbohydr. Res.,* **2018**, *466*, 30-38.
       [http://dx.doi.org/10.1016/j.carres.2018.02.010] [PMID: 29499901]

[33]   Frédéric, C.J.M.; Tikad, A.; Fu, J.; Pan, W.; Zheng, R.B.; Koizumi, A.; Xue, X.; Lowary, T.L.; Vincent, S.P. Synthesis of unprecedented sulfonylated phosphono- *exo* -glycals designed as inhibitors of the three mycobacterial galactofuranose processing enzymes. *Chemistry,* **2016**, *22*(44), 15913-15920.
       [http://dx.doi.org/10.1002/chem.201603161] [PMID: 27628709]

[34]   Habib, S.; Gueyrard, D. Modified julia olefination on sugar-derived lactones: synthesis of difluoro-*exo* -glycals. *Eur. J. Org. Chem.,* **2015**, *2015*(4), 871-875.
       [http://dx.doi.org/10.1002/ejoc.201403175]

[35]   Liu, X.; Yin, Q.; Yin, J.; Chen, G.; Wang, X.; You, Q.D.; Chen, Y.L.; Xiong, B.; Shen, J. Highly stereoselective nucleophilic addition of difluoromethyl-2-pyridyl sulfone to sugar lactones and efficient synthesis of fluorinated 2-ketoses. *Eur. J. Org. Chem.,* **2014**, *2014*(28), 6150-6154.
       [http://dx.doi.org/10.1002/ejoc.201402757]

CHAPTER 2

Recent Advances in the Synthesis of *O*- and *S*-Glycosides from Glycals

Mittali Maheshwari[1], Bindu Tiwari[1], Manish Kumar Sharma[1] and Nazar Hussain[1,*]

[1] Department of Medicinal Chemistry, Institute of Medical Sciences, Banaras Hindu University, Varanasi-221005, India

Abstract: Approximately 20 to 30 percent of all natural drugs are glycosylated, and the attached carbohydrate unit is pivotal for their biological activity. *O*- and *S*-glycosides constitute a significant portion of various biologically potent molecules and natural products. The synthesis of *O*- and *S*-glycosides holds immense importance due to several medicinal benefits. This chapter explores several methodologies for accessing *O*- and *S*-glycosides and the subsequent utilization of these glycosides for oligosaccharide generation. Specific pathways involved in *O*- and *S*-glycosides synthesis are discussed, along with detailed consideration of factors influencing the stereoselectivity of the desired products.

Keywords: Acceptors, Glycosyl donors, Glycals, Oligosaccharides, *O*-glycosides, *S*-glycosides, Stereoisomers.

INTRODUCTION TO *O*-GLYCOSIDES

When one carbohydrate unit is attached to an aglycone or another carbohydrate unit to generate a linkage, it is called a glycosidic linkage [1], and the process is referred to as glycosylation. This linkage can be facilitated through *O*-, *N*-, *S*-, or *C*-glycosidic bonds. *O*-glycosylation involves the coupling of a glycosyl donor with glycosyl acceptors in the presence of promoters and solvent, as depicted in (Scheme **1**) [2]. *O*-glycosides are commonly found in plants [3]. Glycosyl amines or nucleosides contain *N*-glycosides. Glucosinolates or thioglycosides comprise *S*-glycosides. *C*-glycosides are stable glycosides and resistant to hydrolysis due to the existence of a covalent bond between glycone and acceptors [4, 5]. In this process, the glycone acts as the donor molecule, offering its anomeric position, while the aglycone serves as the acceptor molecule. The resulting glycosidic bond

---

* **Corresponding author Nazar Hussain:** Department of Medicinal Chemistry, Institute of Medical Sciences, Banaras Hindu University, Varanasi-221005, India; E-mail: nazar10@bhu.ac.in

Nazar Hussain & Atul Kumar (Eds.)

can manifest as either an α- or β-diastereoisomer, known as α- or β-glycosides, respectively. β-linked *O*-glycosides are prevalent in plant glycosides [6].

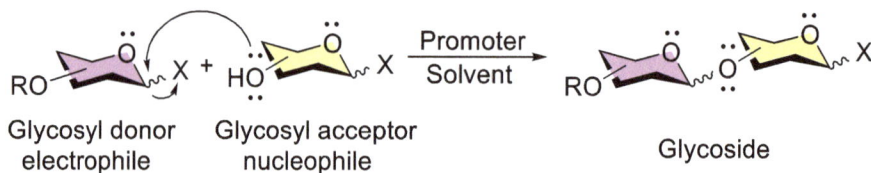

| Glycosyl donor electrophile | Glycosyl acceptor nucleophile | | Glycoside |

Scheme 1. General strategy for the synthesis of *O*-glycosides.

Various novel glycosyl donors have been developed and utilized in the glycosylation process [7 - 10]. (Scheme **2**) illustrates some of the reported glycosyl donors. In earlier times, glycosyl chlorides/ bromides [11] and 1-*O*-hydroxyl sugars [12] were used as donor molecules for creating the glycosidic linkage. Subsequently, during the medieval period, glycosyl fluorides [13], thioglycosides [14], and trichloroimidate [15] were synthesized for the glycosylation purpose. In recent years, a variety of donors have been introduced and employed for the production of *O*-glycosides. Among these, glycals stand out as significant donor molecules [16]. Glycals are derivatives of monosaccharides with an endocyclic double bond at the C1/C2 position, serving as versatile building blocks in the assembly of natural products and biologically active molecules [17]. Essentially highly substituted cyclic vinylic enol ethers, glycals exhibit distinct reactivity at the C1 and C2 positions due to the conjugation of the ring oxygen with the double bond [18]. Recent advancements in glycals and their transformation into diverse biologically active molecules have broadened their application in modern organic synthesis. Since their initial discovery and synthesis by Fischer and Zach in 1913, glycals have been extensively utilized as pivotal precursors for synthesizing numerous biologically significant molecules [19]. The extensive transformation of glycals into various biologically active molecules and natural products stems from their unsaturation and facile conversion into diverse derivatives and precursors for further derivatization [20].

Numerous strategies have been proposed for the glycosylation of glycals with various nucleophiles, leading to the formation of *O*-glycosides [21 - 23]. In this chapter, we will discuss the various approaches employed to activate donors effectively for the synthesis of *O*-glycosides and apply these protocols in achieving high yields in oligosaccharide synthesis. Our focus primarily lies in the methodologies introduced since 2017.

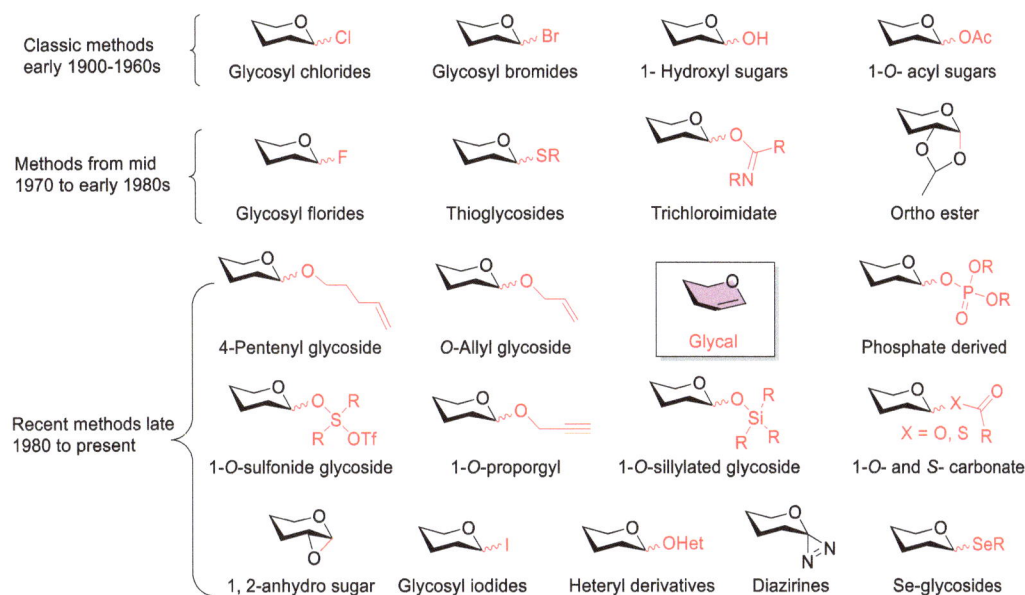

Scheme 2. Different donor molecules used for glycosylation.

HISTORICAL ASPECTS

In the late 19th century, Michael [24] and Fischer [25] presented the chemical synthesis of the first *O*-glycosides, namely *p*-methoxyphenyl β-D-glucopyranoside and methyl α-D-glucopyranoside, respectively. Subsequently, Koenigs and Knorr [11] proposed modified and controlled protocols for *O*-glycoside synthesis. Over time, various methodologies have been developed, advancing mechanistic pathways and enhancing control over the stereochemistry of the desired products. Glycosylation can yield two different anomeric stereoisomers: 1,2 *cis*- and 1,2 *trans*-isomers. Lemieux *et al*. [26] and Ness [27] emphasized the significance of the protecting group at the C2 position of glycosyl donors on product stereochemistry. Fraser-Reid *et al*. [28] introduced the concept of armed and disarmed approaches, where an ether linkage at the C2 position (armed donor) results in 1,2 *cis*-glycosides, while an ester bond (disarmed donor) yields 1,2 *trans*-products.

To optimize glycosylation conditions, various modified approaches have been developed [29]. One-pot glycosylation has gained popularity, incorporating numerous advancements and novel reaction conditions [30 - 32]. Computational chemistry and kinetic studies have contributed to elucidating the mechanism of glycosylation. Despite efforts in this branch of carbohydrate chemistry, mastering the mechanism remains a challenge.

Medicinal Applications of O-glycosides

O-glycosides are ubiquitous structural motifs present in a wide range of natural products and commercially available drug molecules [23]. (Scheme **3**) represents some naturally occurring *O*-glycoside drugs having various biological activities. They include Digoxin [33], which is used to treat congestive heart failure and heart rhythm problems called atrial fibrillation; Oleandrin [34], which is a lipid-soluble cardiac glycoside isolated from the plant Nerium oleander and utilized to treat congestive heart failure; Quabain [35], which is a cardiac glycoside used to treat hypotension and some arrhythmias; Lomaiviticin A [36], which acts as an antibiotic and drug obtained from micromonospora lomaivitiensis; Erythromycin A [37], which is utilized to prevent and treat different infections like respiratory tract infections, skin infections, *etc.*; Tylosin [38], which acts as an antibiotic drug and belongs to the same family as erythromycin; Mycinamicin [39], which acts as an antibiotic drug; Menomycin A [40], which is used to treat acne, which is resistant to other antibiotics; Landomycin [41], which is used as an antibiotic and anticancer drug; Streptomycin [42], which acts as an aminoglycoside antibiotic and is used to treat certain kinds of bacterial infections.

Mechanistic Insight into *O*-glycosylation

The general mechanism of *O*-glycosylation [43] begins with the activation of the leaving group of glycosyl donor **1** in the presence of a promoter, as illustrated in (Scheme **4**). This activation leads to the formation of oxocarbenium ion **1A**, adopting a flattened half-chair conformation. Subsequently, nucleophilic acceptors attack **1A**, resulting in the formation of the desired glycosides **3**. The attack of the acceptors can occur *via* two pathways, depending on the structural properties of the oxocarbenium intermediate. If the nucleophiles attack from below **1A**, α-glycosides **3a** are formed, and if the nucleophiles attack from above **1A**, β-glycosides **3b** are generated.

Factors Affecting the Stereochemistry of the Glycosides

The stereochemistry of glycosides primarily depends on three major factors: 1) neighboring group participation, 2) anomeric effect, and 3) solvent effect.

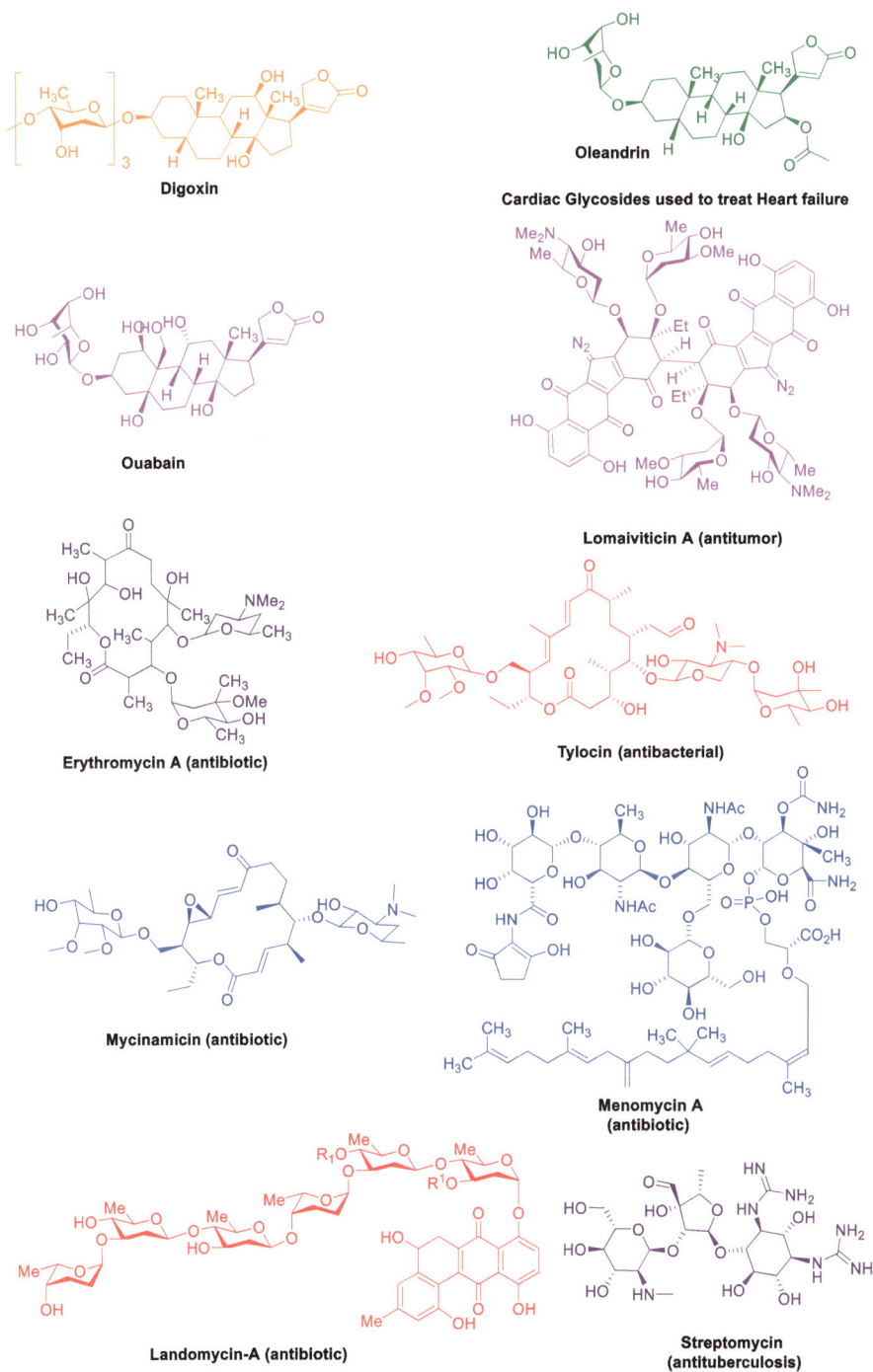

Digoxin

Oleandrin

Cardiac Glycosides used to treat Heart failure

Ouabain

Lomaiviticin A (antitumor)

Erythromycin A (antibiotic)

Tylocin (antibacterial)

Mycinamicin (antibiotic)

Menomycin A (antibiotic)

Landomycin-A (antibiotic)

Streptomycin (antituberculosis)

Scheme 3. General medicinal application of *O*-glycosides.

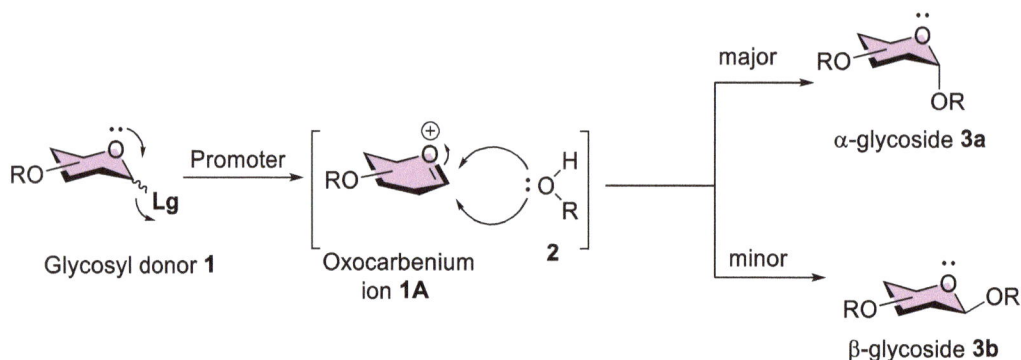

Scheme 4. General mechanistic pathways.

Neighboring Group Participation

Neighboring groups [44, 45] play a significant role in the reaction mechanism, dictating the stereochemistry of the desired product, as depicted in (Scheme 5). When an acyl group is situated at the C2 position of the glycosyl donor, it interacts with oxocarbenium ion **4A** to form acyloxocarbenium **4B**. Subsequently, the nucleophile primarily attacks from above the intermediate, leading to the formation of β-glycosides or 1,2 *trans*-glycosides **5b**.

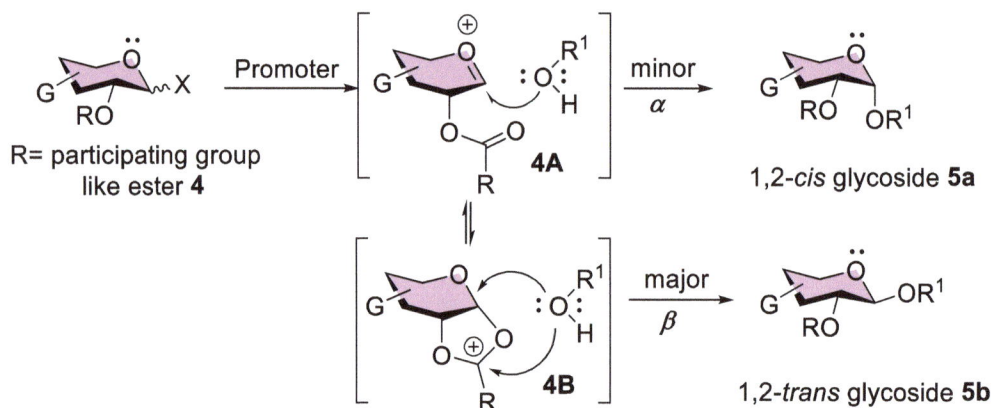

Scheme 5. Mechanistic pathway involving NGP.

Anomeric Effect

Edward and Lemieux, between 1955 and 1988, first defined the anomeric effect, as depicted in (Scheme 6). The anomeric effect, also called the Edward-Lemieux effect, depends on the stereochemistry of the C1 position of the pyranose ring [30, 46, 47]. The anomeric effect involves two factors: the stereoelectronic effect and the dipole-dipole interaction. According to the stereoelectronic effect, the α isomer is more stable due to the hyperconjugation of the nonbonding electron pair of oxygen atoms with the vacant σ^* orbital of the C–X bond, where α stabilizes the axial configuration. Dipole-dipole interaction states that in β-isomer, the dipoles of the heteroatom are partially aligned and therefore repel each other. But in α-isomer, the opposite orientation of the dipole stabilizes the system and shows that the α-isomer is a more stable product.

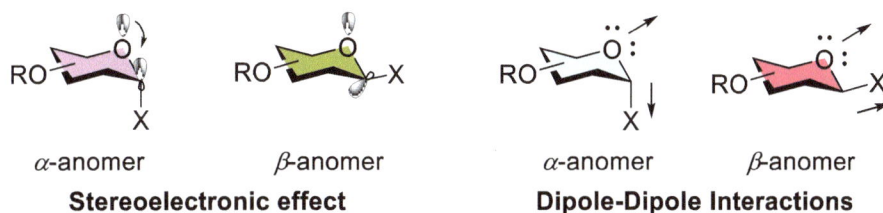

α-anomer	β-anomer		α-anomer	β-anomer
Stereoelectronic effect			**Dipole-Dipole Interactions**	

Scheme 6. Representation of anomeric effect.

Solvent Effect

The stereoselectivity of glycosides is also influenced by the solvent used in the reaction, as illustrated in (Scheme **7**) [48, 49]. In an ether solvent, α-glycosides are formed, whereas in acetonitrile, β-glycosides are synthesized (Scheme **7B**). In acetonitrile, CH_3CN initially binds to the oxocarbenium ion at the α position of the ring. This blocks the α position, causing nucleophilic attack to occur from the β-side, resulting in the production of β-glycosides. In ether (Scheme **7A**), the solvent molecule coordinates at the β-position of the ring, allowing the acceptor to attack from the α side and generate α-glycosides.

Literature on Synthesizing O-glycosides

Here, we have included some recent research articles showing the preparation of *O*-glycosides, including 2-deoxy *O*-glycosides from glycosyl donors. Some literature reports utilizing *O*-glycosides for further synthesis of oligosaccharides are also discussed in this chapter.

In 2014, M. C. Galan and their coworkers [50] developed a method to prepare glycoconjugates or disaccharides from glucals and rhamnals in high yield with α-selectivity, as depicted in (Scheme **8**). The protocol utilized *trans*-fused cyclic 3, 4-*O*-disiloxane protecting group and TsOH·H$_2$O as a catalyst for the conversion of glycals to glycosides. Substrate scope was explored by using different donors with 6-OH saccharides, resulting in the formation of disaccharides in good yield.

A)

Et-O-Et

ROH

α–glycosylation

B)

CH$_3$CN

ROH

β–glycosylation

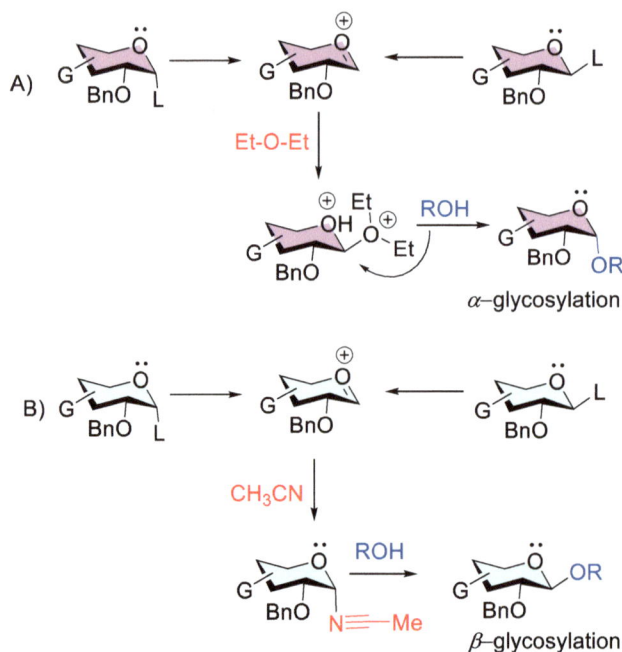

Scheme 7. Representation of solvent effect.

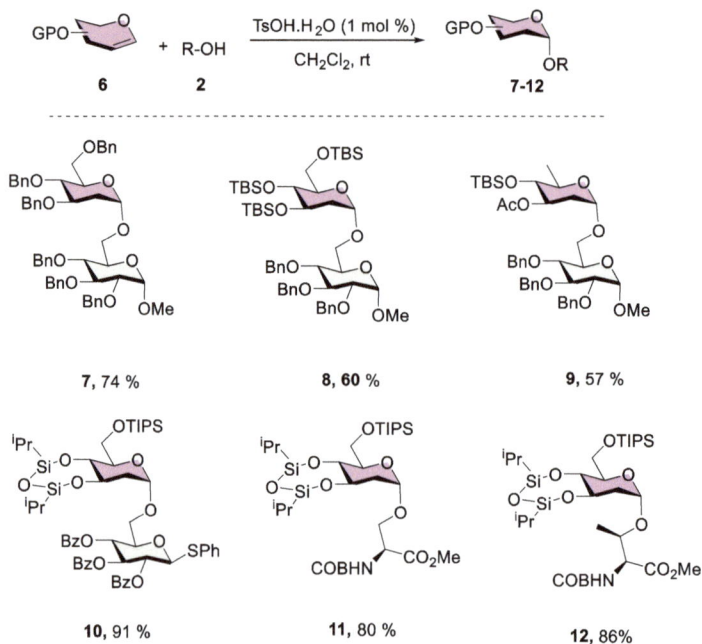

GPO + R-OH $\xrightarrow[\text{CH}_2\text{Cl}_2,\ \text{rt}]{\text{TsOH.H}_2\text{O (1 mol \%)}}$ GPO—OR

6   **2**   **7-12**

7, 74 %   **8, 60** %   **9**, 57 %

10, 91 %   **11**, 80 %   **12**, 86%

Scheme 8. A 3,4-*trans*-fused cyclic protecting group facilitates selective catalytic synthesis of 2-deoxyglycosides.

Perbenzylated glucals react with the nucleophiles, affording disaccharides in 4 h with 74% of yield (**7**). Persilylatedglycal reacts under optimized conditions, resulting in the formation of disaccharides in 60% yield (**8**). After that, rhamnal binds with the saccharides, giving a desired product with a 57% yield. When 3,4-*O*-tetramethyldisiloxane is used as a donor, it reacts with different acceptors to produce different glycomimics in high yield (**9-12**).

In 2014, M. Tanaka and coworkers presented regio and stereo-controlled, one pot amido glycosylation of glycals in the presence of rhodium catalyst, as shown in (Scheme **9**) [51].

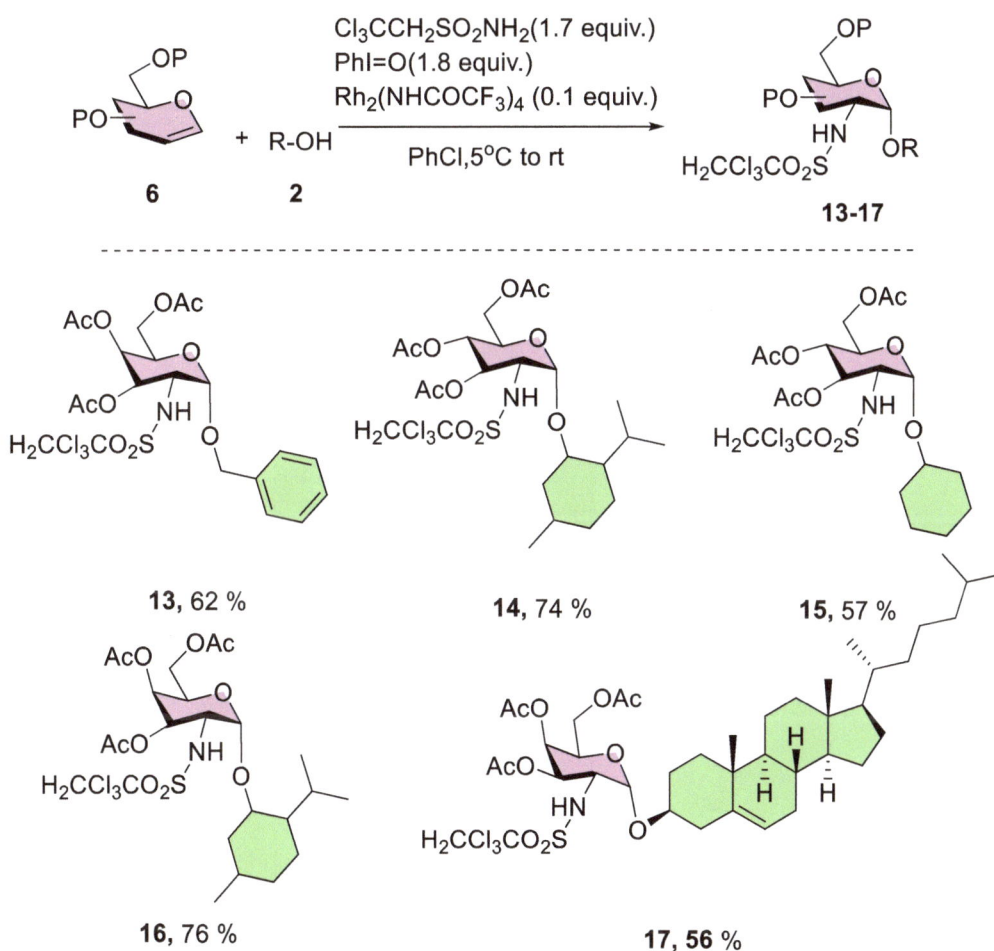

Scheme 9. Stereo-controlled amidoglycosylation of alcohols with acetylated glycals and sulfamate ester.

The generality and scope of the reaction were investigated by reacting glycals with different acceptors, resulting in the formation of aminoglycosides in good yield. Primary alcohols **13** and secondary alcohols **14-16** undergo reaction with glucal and galactal, giving the desired product in good yield. Cholesterol also reacts with glucal to give aminoglycosides in 56% yield **17**.

In 2014, Y. D. Vanker and coworkers [52] developed a strategy involving gold chloride and phenylacetylene to promote Ferrier glycosylation with different nucleophiles (Scheme **10**). Varieties of glycosides were prepared by reacting glucal with different nucleophiles such as benzyl alcohol, cyclohexanol, and 6-OH glucosides (**19-21**). Substrate scope was also prepared from galactal as a donor with different nucleophiles, resulting in the formation of galactosides (**22-24**).

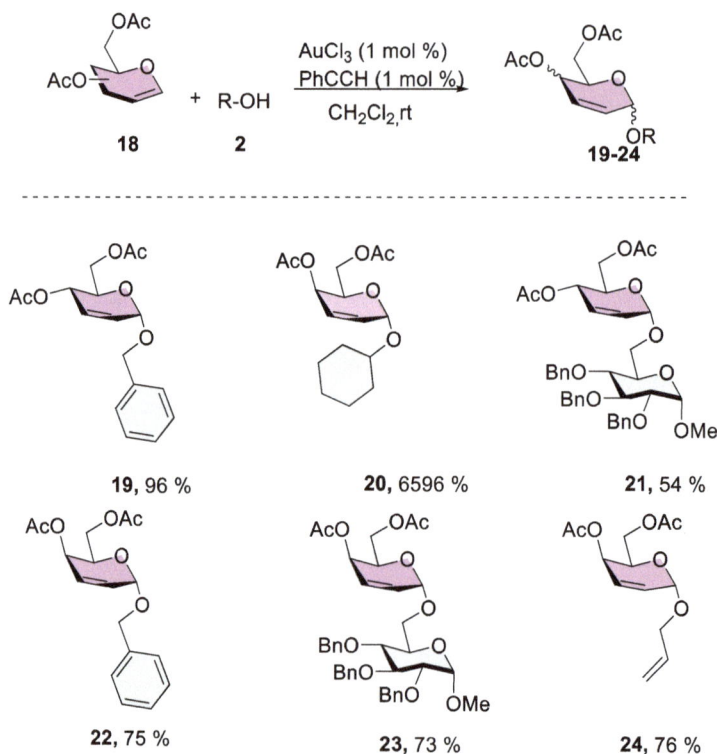

Scheme 10. Gold (III) chloride and phenylacetylene catalyzed Ferrier rearrangement.

In 2015, J. Su and coworkers [53] performed Ferrier glycosylation by using gadolinium triflate, as shown in (Scheme **11**). A series of *O*-pyranosides were prepared by Ferrier glycosylation with different nucleophiles. Initially, different acceptors bind with acetyl D-glucal to give glycosides in good yield (**25-27, 19**). Then, glycosides are formed by variation in sugar derivatives such as with 3,4,6-tri-*O*-benzyl-D-glucal **29** and 3,4-di-*O*-acetyl-L-rhamnal **28** under optimized conditions to yield the desired product in excellent yield.

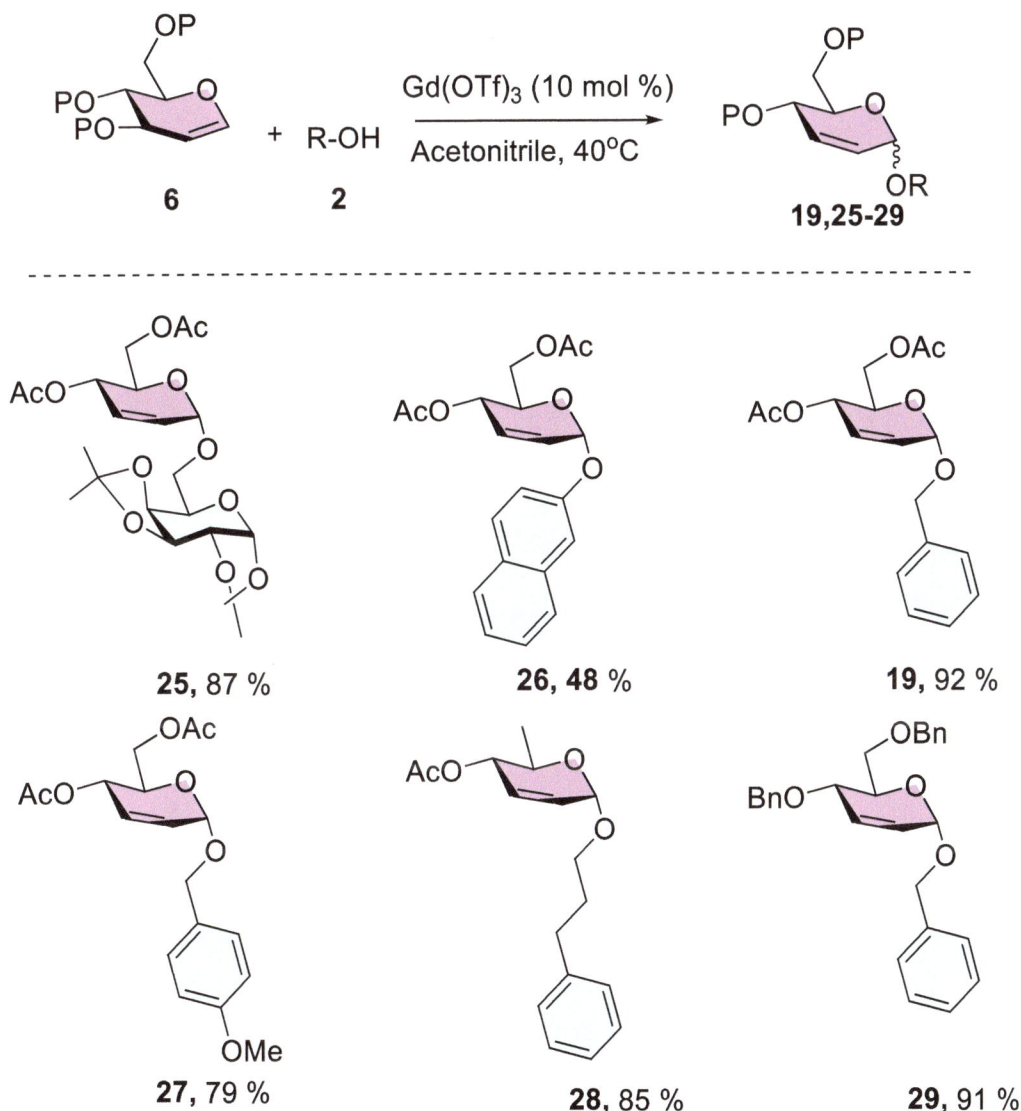

Scheme 11. Gd(OTf)$_3$-catalyzed preparation of 2,3-unsaturated *O*-pyranosides from glycals by Ferrier rearrangement.

In 2016, V. H. Jadhav and coworkers [54] derived the magnetic solid acid catalyst for facile glycosylation of glycals, as depicted in (Scheme **12**). This new porous magnetic carbonaceous solid acid catalyst Glu-Fe$_3$O$_4$-SO$_3$H was characterized by using different techniques like PXRD, FT-IR, SEM, XPS, *etc.* Substrate scope was explored with different nucleophiles, such as propargyl alcohol **32**, octanol **33**, and methanol **34**, affording the desired product in good yield.

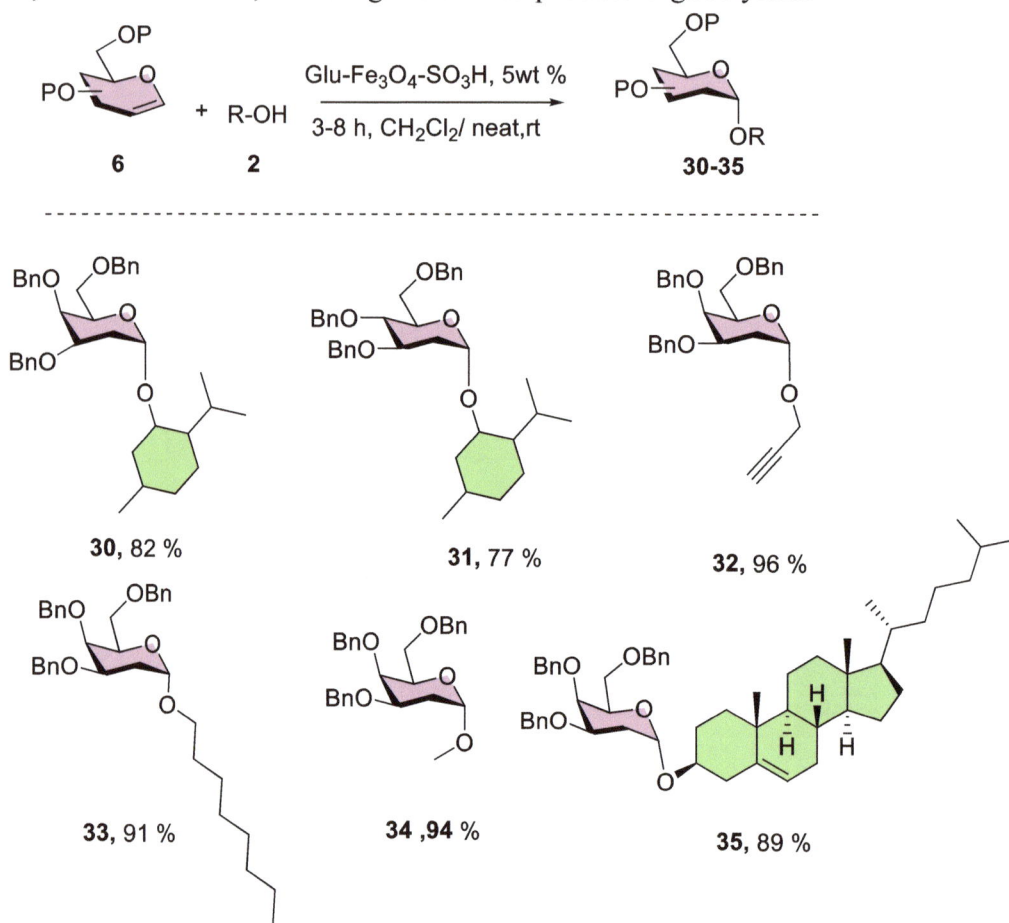

Scheme 12. Facile *O*-glycosylation of glycals using Glu-Fe$_3$O$_4$-SO$_3$H, a magnetic solid acid catalyst.

Different sugars were also reacted with menthols to give an excellent yield of the product **30-1**. Cholesterol was also reacted with enolic ethers, resulting in glycosides **35** in 89% yield. The catalyst can also be separated after reaction with an external magnetic field and can be reused up to 4 cycles without decrement in yield.

In 2017, M. C. Galan and coworkers [55] demonstrated that Pd(MeCN)$_2$Cl$_2$ catalyzed α-stereoselective synthesis of 2,3-unsaturated *O*-glycosides from acetyl-protected glycals (Scheme **13**). A series of differently protected glycosides were synthesized in good to excellent yield. Different sugars like acetyl-protected glucal **25**, acetylated galactal **36**, and conformationally constrained glucal **37** reacted to give disaccharides in moderate to good yield with high selectivity. Primary alcohols undergo glycosylations with D-glucal to afford 2,3-unsaturated glycosides (**38-39**) in 90-97% (9:1 α:β ratio) yield within 3-7 h. Glycosylation of secondary alcohols required more time (15-17 h), and additionally, in the case of *Boc*-protected serine, it required a long time and a high-temperature range (50 °C) to yield 68% of the desired product (**40**).

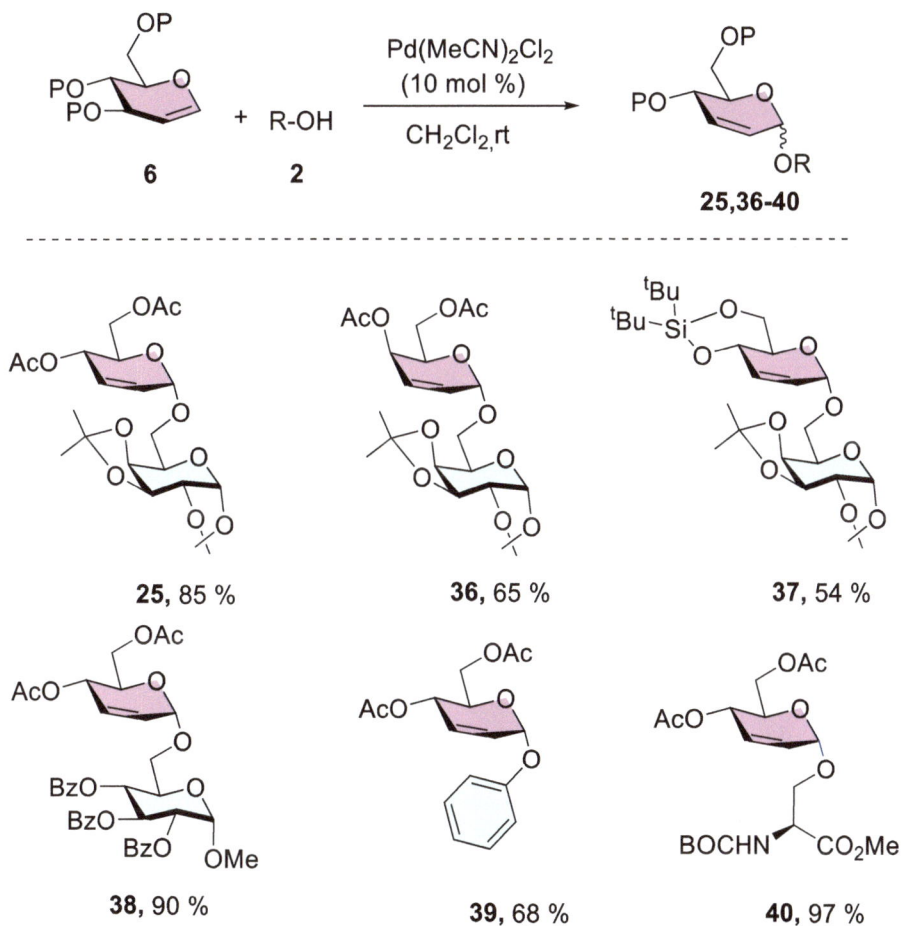

Scheme 13. Palladium-catalyzed α-stereoselective *O*-glycosylation of (3)-acylated glycals.

In 2017, M. C. Galan and coworkers [56] presented the direct and stereoselective synthesis of deoxy *O*-glycosides from glycals, as depicted in (Scheme **14**). The combination of Au (II) and silver triflate was utilized at room temperature for the functionalization of enolic ethers. To demonstrate the tolerance of the catalytic system, the substrate scope was explored with different alcohols. By reacting with simple primary alcohols such as benzyl alcohols **44**, glycosides **42** and **43** give a smooth reaction with good α-selectectivity and yield. Reaction with secondary alcohols such as *N*-hydroxy succinimide **45**, *Boc*-protected threonine **46**, and (R)-(-)-1-(2-naphtyl) ethanol **47** also produce the desired product with good yields ranging from 65-85%.

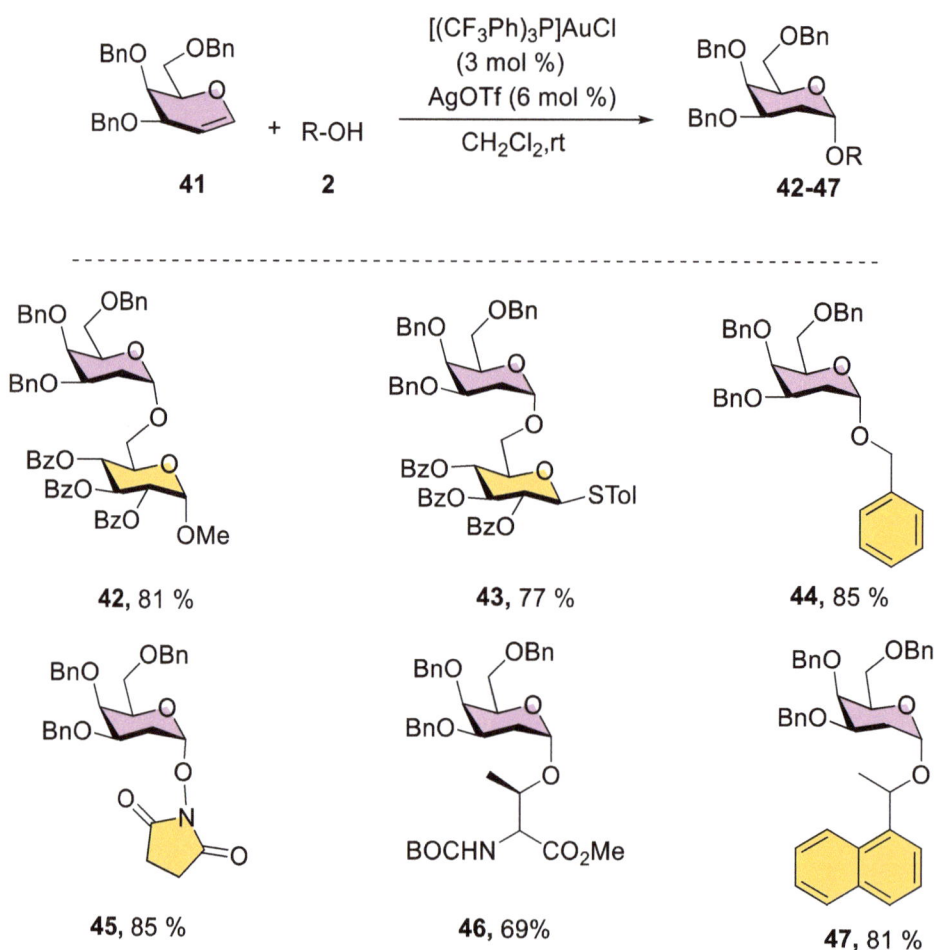

Scheme 14. Gold(I)-catalyzed direct stereoselective synthesis of deoxyglycosides from glycals.

In 2017, X. W. Liu and coworkers [57] constructed numerous *O*-glycosides from 3,4-*O*-carbonate galactal as a donor molecule using palladium catalysis, as depicted in (Scheme **15**). The glycal donor first coordinated with the palladium catalyst from the β-face directed by the carbonate group. After that, aliphatic alcohols (hard nucleophiles) **49-51** give β-glycosides, and phenols (soft nucleophiles) **53-54** give α-glycosides. During the screening of ligands, various ligands were optimized, but xantphos facilitated the reaction in 1 h as it is a bidentate ligand with a suitable bite angle.

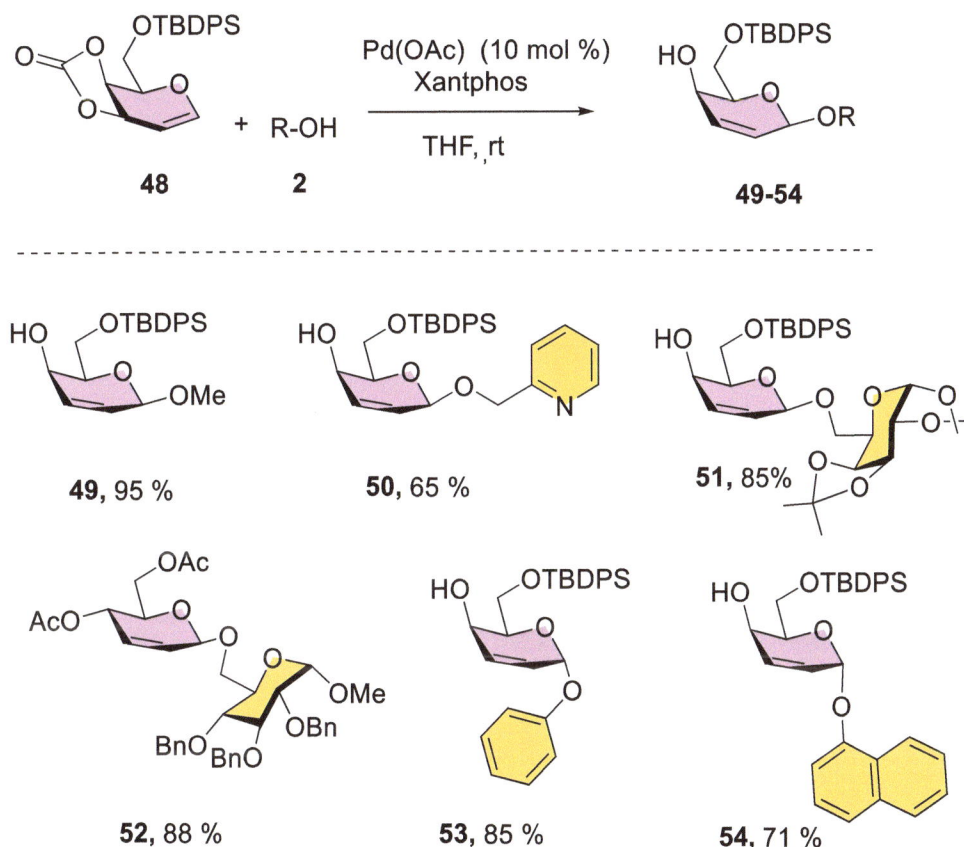

Scheme 15. Catalyst-controlled stereoselective *O*-glycosylation.

In 2017, N. L. B Pohl and coworkers [58] proposed the glycosylation of *S*-containing glycosyl donors by using triphenyl bismuth ditriflate, propanethiol, as an additive for 3 h, as depicted in (Scheme **16**). The protocol was utilized to activate uronic esters of *S*-Phenyl, *S*-adamantyl, *S*-benzoxazolyl, *S*-thiazolinyl glycosides, and sialic acid. Glycosides were synthesized by using fluorous-tagged alcohol acceptors, and it was observed that with the addition of an acceptor solution and propanethiol additive, the color of the reaction immediately changed. This novel pentavalent bismuth activation proved to be successful in activating thioglycosides to afford *O*-glycosides with 61-85% yield. The reaction of glucosides **58-59** and galactosides **56-57** resulted in the formation of the desired product in good yield.

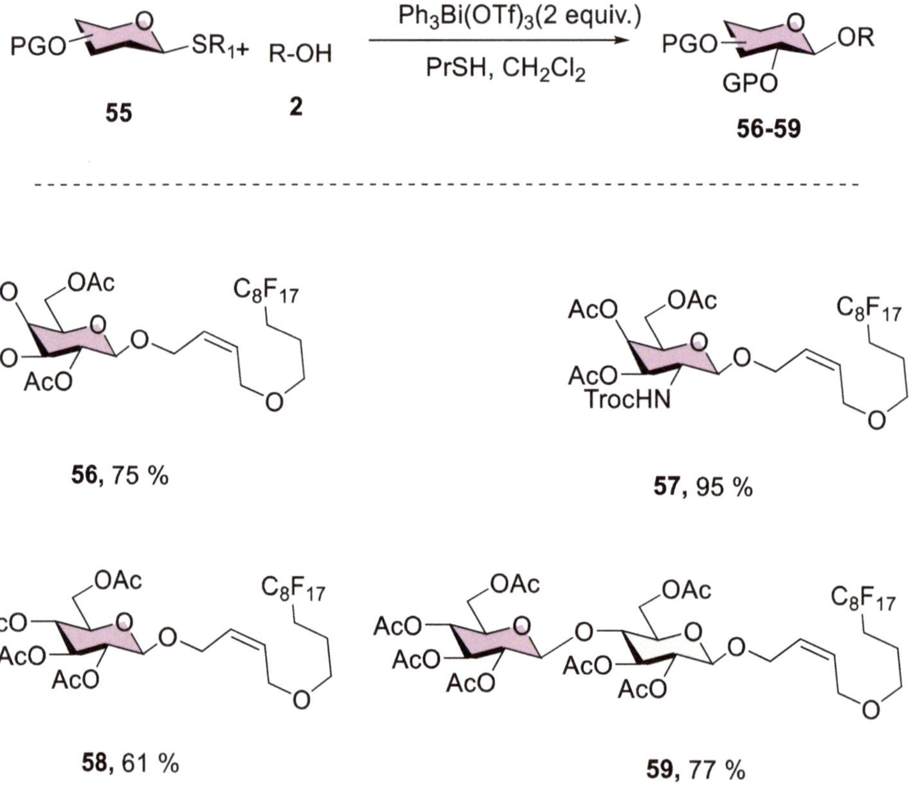

Scheme 16. Pentavalent bismuth as a universal promoter for *S*-containing glycosyl donors with a thiol additive.

In 2018, K. Toshima and coworkers [59] presented 1,2-*cis*-stereoselective and regioselective glycosylation of glycal donors to derive natural drugs, as shown in (Scheme **17**). 1,2-*cis*-glycosylation was carried out naturally *via* glycosyl transferases enzyme through the SNi mechanism. The optimized reaction involved p-amino boronic acid as a catalyst in acetonitrile with water to produce desired glycosides in good yield. The scope of the reaction was explored by using

1,2-anhydro sugars and unprotected acceptors. 1,2-anhydro sugar glycosylation gave 1,4 glycosides **61-63**, 1,6 mannosides **62**, and 1,4 galactosides **64**. Trissacharides **65** were also prepared by using the optimized methodology to enhance the generality of the reaction condition in 60% yield.

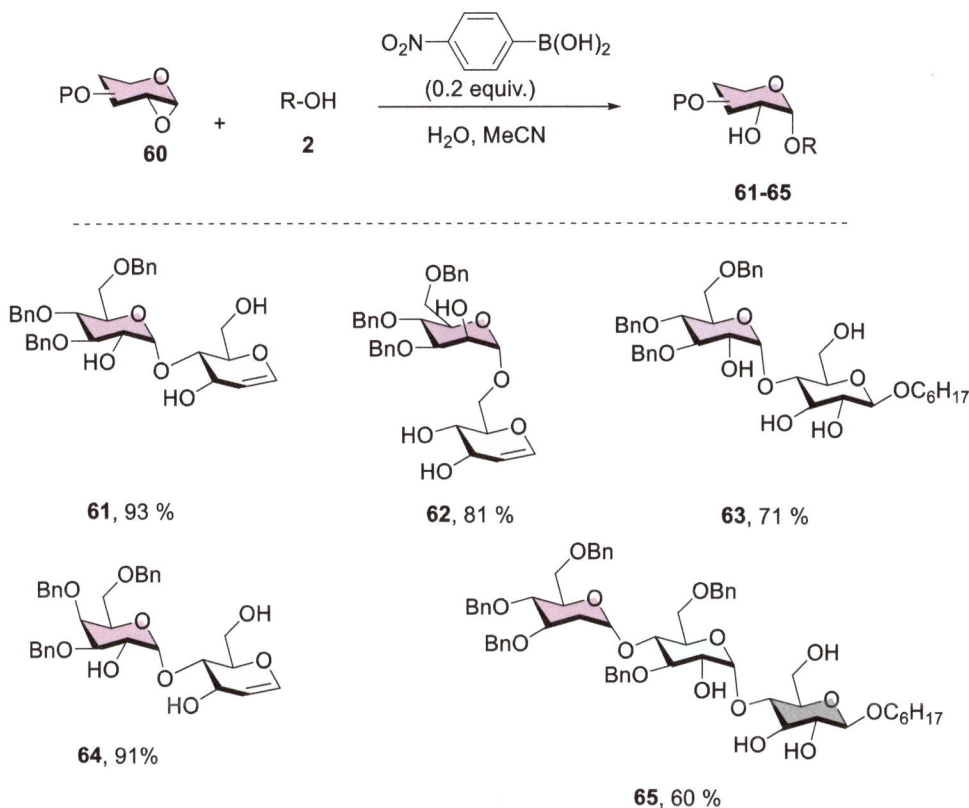

Scheme 17. Boronic acid-catalyzed regioselective and 1,2-*cis*-stereoselective glycosylation of unprotected sugar acceptors *via* SNi-type mechanism.

In 2018, M. A. Walczak and coworkers [60] described the oxidative aerobic glycosylation of glycosyl stannanes **20** with nucleophiles to achieve β-glycoside in good to excellent yield, as shown in (Scheme **18**). These reactions were promoted in the presence of hypervalent iodine and a catalytic amount of zinc salt. The reaction was performed with different alcohol acceptors, resulting in the formation of β-glycosides without the formation of ketone or aldehyde as a byproduct. Different glycosyl stannanes were utilized, like glucose **67**, galactose **68**, and fucose **69**, to give β-glycosides. The protocol was further used to synthesize diasaccharide **70** and oligosaccharide **71** in good yield.

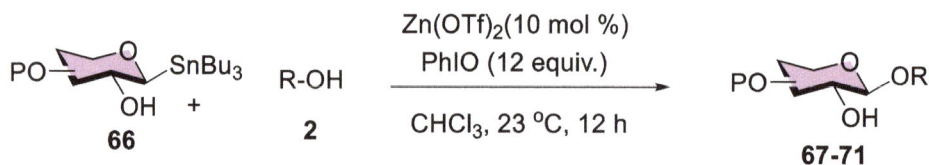

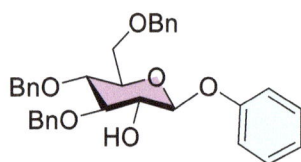
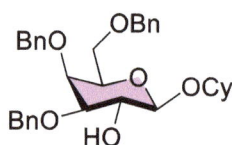
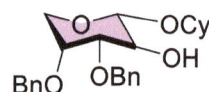

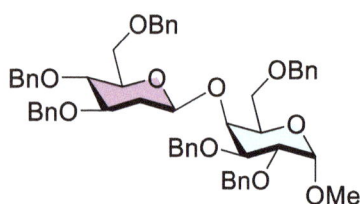

67, 61 %   **68**, 72 %   **69**, 57 %

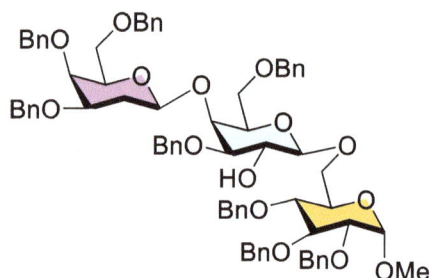

70, 73%   **71**, 63 %

Scheme 18. Stereoselective oxidative glycosylation of anomeric nucleophiles with alcohols.

In 2019, M. Xia and coworkers [61] proposed that perfluorophenylboronic acid catalyzed the direct stereoselective synthesis of 2-deoxygalactosides in 55-88% yield with complete α-selectivity (Scheme **19**). A series of disaccharides were synthesized from peracetylated D-galactal in moderate yield (**72-75**). The glycosylation also works on a broad range of acceptors and tolerates wide functional groups, resulting in the desired glycosides in good yield without the formation of Ferrier products as a side product (**75-77**).

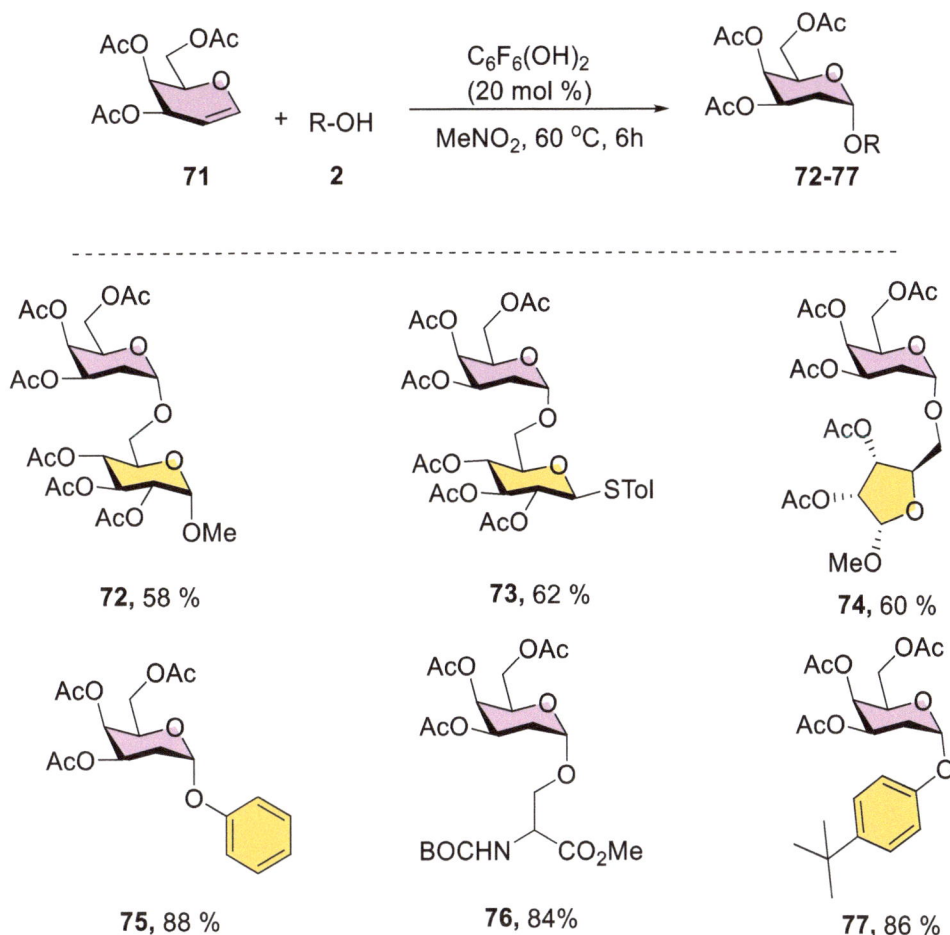

Scheme 19. Perfluorophenylboronic acid-catalyzed direct-stereoselective synthesis of 2-deoxygalactosides from deactivated peracetylated D-galactal.

In 2019, Y. Chai and coworkers [62] revealed the chemoselective glycosylation of rhamnal donors with alcohol acceptors, as well as access to rhamnosides in good yield and high α-selectivity (Scheme **20**). Reaction in a kinetically controlled pathway gives 2-deoxy rhamnosides, and in a thermodynamically controlled pathway at elevated temperature, it yields 2,3-unsaturated rhamnosides. Silyl-protected rhamnal donors react with primary alcohols and secondary alcohols, generating 2-deoxy rhamnoside in good yield at room temperature (**79-81**). When the same reaction with aglycons takes place at elevated temperatures, it produces a Ferrier product with a high yield (**82-84**).

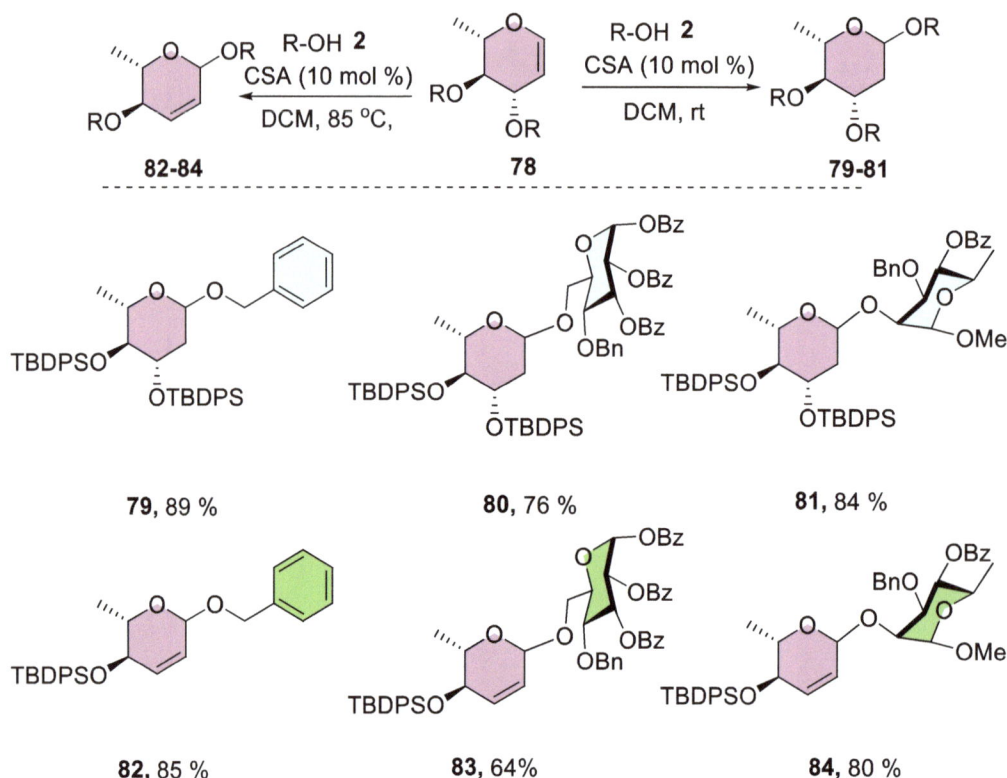

Scheme 20. Tuning the chemoselectivity of silyl-protected rhamnals by temperature and brønsted acidity: kinetically controlled 1,2- addition *vs.* thermodynamically controlled Ferrier rearrangement.

In 2020, X. W. Liu and coworkers [63] reported that the superbase catalyzed the glycosylation of 2-nitroglycals to yield 2-amino-2-deoxy-*O*-glycosides **86-91** (Scheme **21**). The ion pair produced from alcohol and a catalytic amount of P_4-t-Bu is necessary for stereo selective glycosylation to give the product in moderate to good yield. 2-nitro galactals, under optimized reaction conditions, produce α-stereoisomers (**86-88**), and 2-nitroglucals yield more β-stereoisomers (**89-91**). The synthetic utility of the protocol was demonstrated by synthesizing a key intermediate of a mucin-type core-6 glycoconjugate.

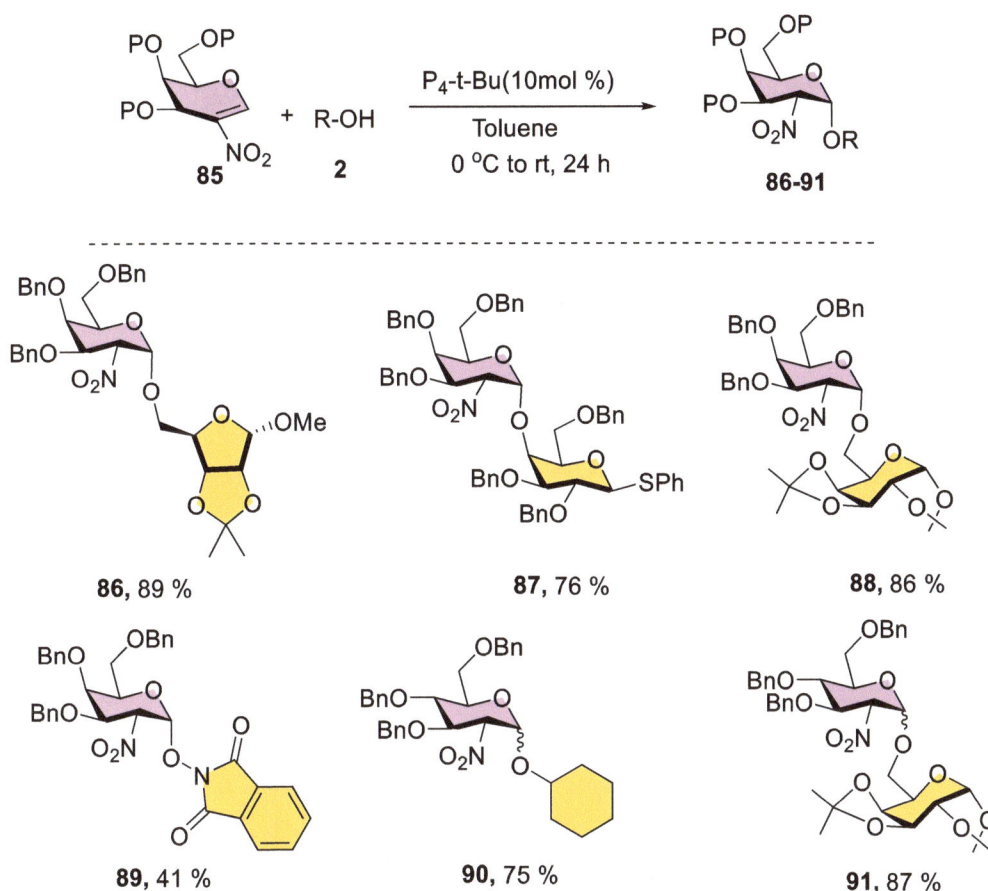

Scheme 21. Superbase catalyzed stereo- and regioselective glycosylation with 2-nitroglycals.

In 2020, M. C. Galan and coworkers [64] presented the activation of both armed and disarmed glycals towards the synthesis of α-stereoselective *O*-glycosides (Scheme **22**). Primary and secondary alcohols were made to react under optimized conditions, proceed smoothly, and give good to excellent yield of the product. Primary alcohols of glucose and galactose were reacted to afford disaccharides in good yields with α-selectivity (**92-94**). Glycosylation with simple benzyl alcohol gave the desired glycoside **95** in 82% yield within 2 h and with >30:1 α:β ratio. Similarly, secondary alcohols like *Boc*-protected threonine **46** and cholesterol **96** yield glycosides with good yield and great α-selectivity.

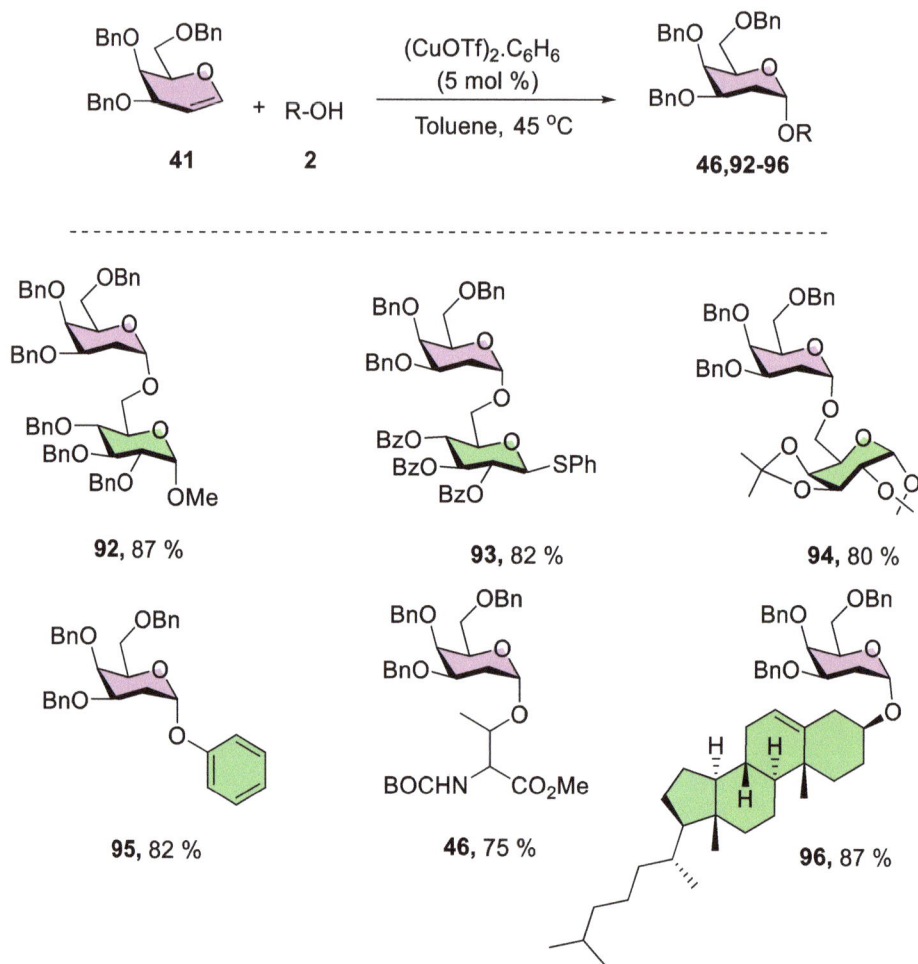

Scheme 22. Copper reactivity can be tuned to catalyze the stereoselective synthesis of 2-deoxyglycosides from glycals.

In 2020, S. Kashyap and coworkers [65] presented the activation of armed and disarmed donors in the presence of copper triflate as a catalyst, as shown in (Scheme **23**). The protocol was carried out under optimized conditions with different nucleophiles to afford desired glycosides. Primary **44, 92, 98-99** and secondary **97, 100** alcohols reacted with benzyl-protected glycals, resulting in the formation of *O*-glycosides in 72-94% yield with high alfa anomeric selectivity. Glycal donors also react with 6-OH and 4-OH glycosides to give disaccharides in excellent yield (**92, 99-100**).

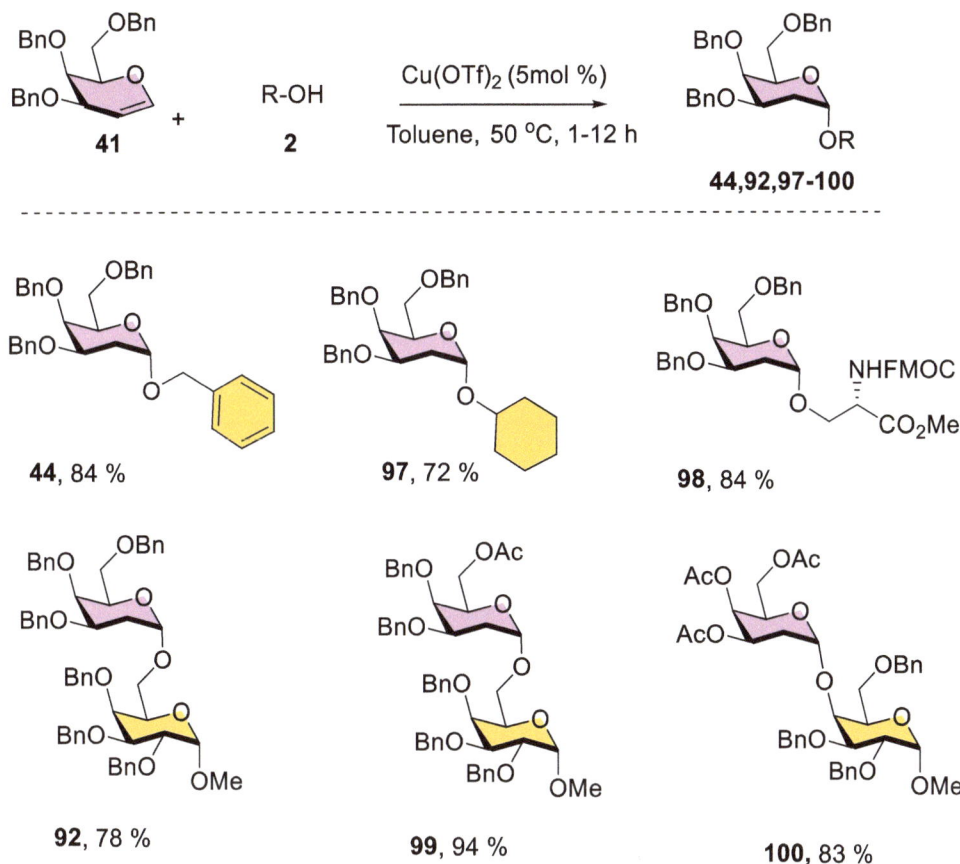

Scheme 23. Copper(II)-catalyzed stereoselective 1,2-addition *vs.* ferrier glycosylation of "armed" and "disarmed" glycal donors.

In 2020, J. Zhang and coworkers [66] proposed an efficient methodology for the synthesis of 2-deoxy *O*-galactosides, as depicted in (Scheme **24**). The protocol utilized 5 mol % of $CuBr_2$ in dichloromethane at room temperature. The substrate scope of the glycosides was explored by using benzyl-protected galactal as a donor molecule with different types of acceptors. Aromatic nucleophiles such as phenol reacted to give galactoside **95** in 80% yield. Aliphatic nucleophiles reacted with galactal, affording the desired product in excellent yield (**44, 93, 94, 101-102**). After that, different protected galactals utilized in the reaction condition, like ethyl protected, reacted with the acceptors, resulting in the formation of disaccharide **102** in 76% yield.

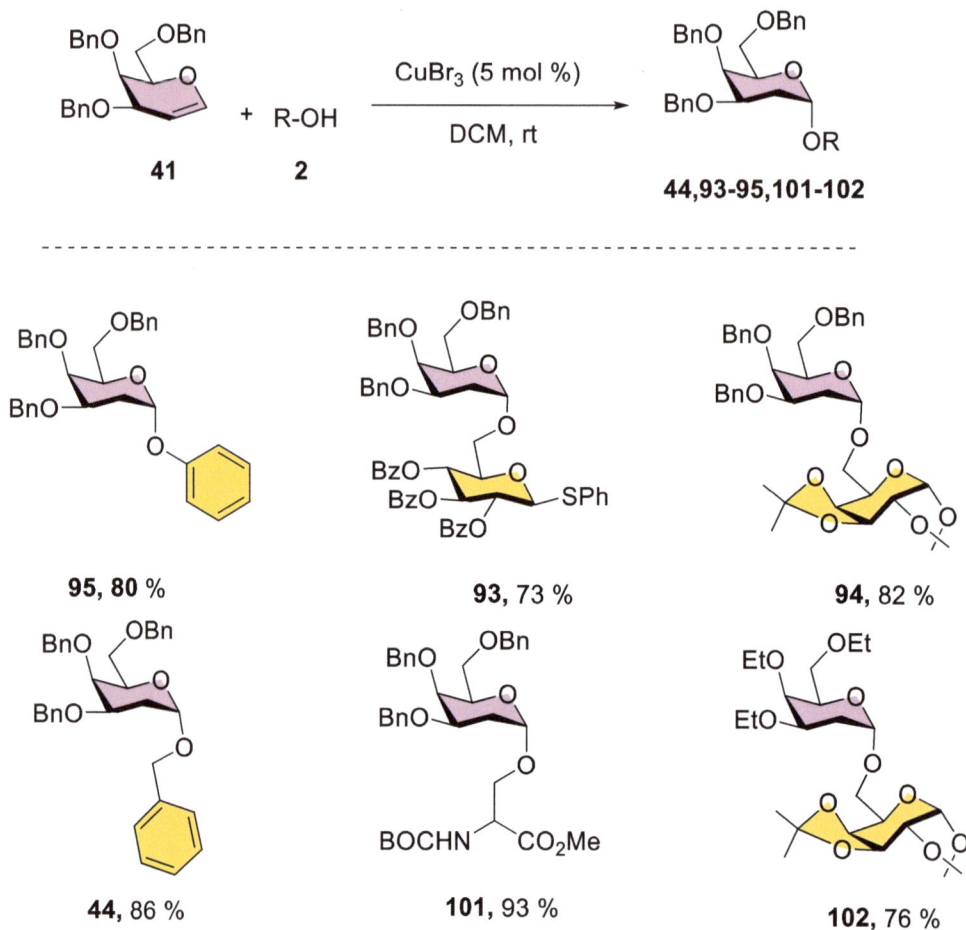

Scheme 24. Copper-catalyzed stereoselective synthesis of 2-deoxygalactosides.

In 2021, X. W. Liu and coworkers [67] demonstrated the stereoselective synthesis of 2-deoxy *O*-glycosides using the iridium complex as a catalyst (Scheme **25**). The donor can be successfully utilized with different acceptors like primary alcohols **103-107**, resulting in the desired product in good yield. Glycosyl donor also binds with N-hydroxyphthalimide to obtain 78% of 2-deoxy glycoside **108**.

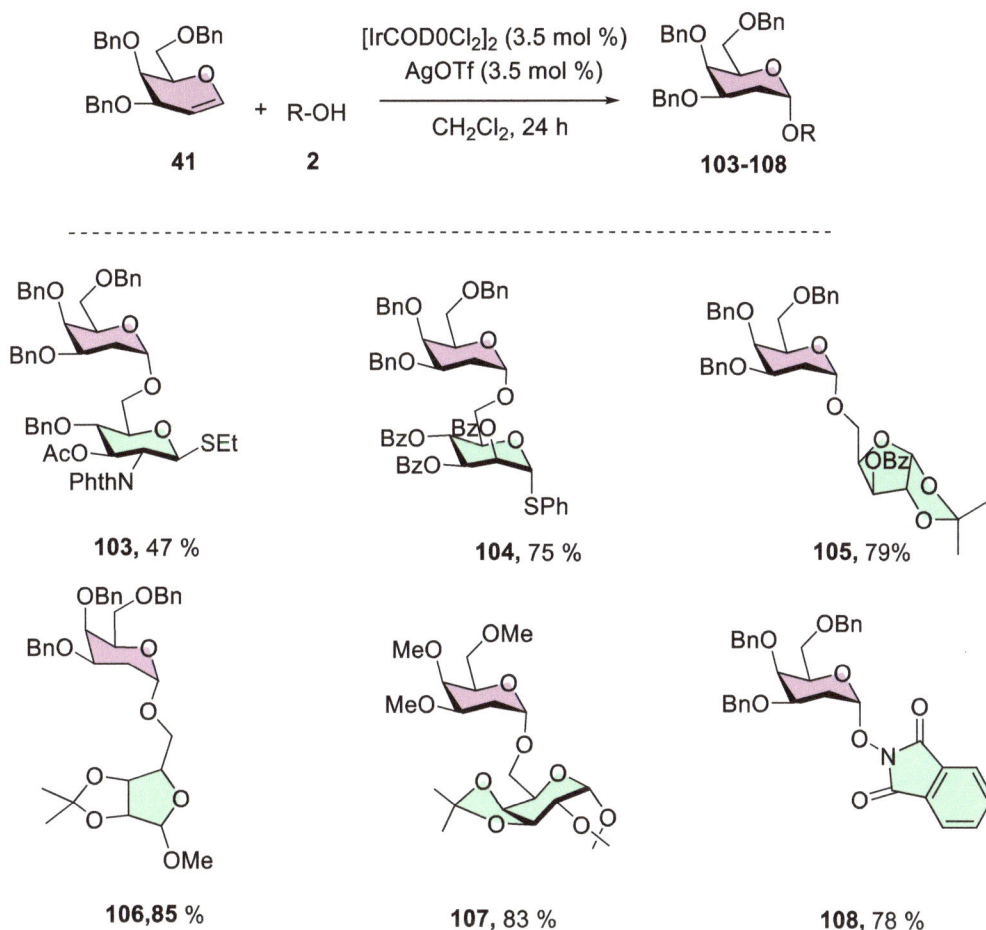

Scheme 25. Iridium-promoted deoxyglycoside synthesis.

In 2021, C. Xia and co-workers [68] proposed a strategy for the synthesis of 1,2-trans-2-amino-2-deoxyglycosides in a single step (Scheme **26**). The amido-glycosylation was initiated by the formation of a benzene sulfonimide radical from NFSI through the reduction of TEMPO. This radical subsequently attacked electrophilically on glycals, generating another radical, which, upon oxidation with TEMPO, produced an oxocarbenium ion. Acceptors then attacked the oxocarbenium ion to yield the β-glycosides. Various glycal donors were investigated in this study, such as benzyl ether **109** and methyl ether **110**, yielding the corresponding disaccharides in good yields (β: $\alpha > 20:1$ or β only). Acetyl-protected glycals **111** provided the desired product with good selectivity but in a lower yield. The use of epiandrosterone as an acceptor in the reaction resulted in the formation of a product with good yield and selectivity **112**. Additionally, a salidroside analog **113**, known for its protective activities against hypoglycemia and serum limitation-induced cell death in rat pheochromocytoma cells, was synthesized with good yield.

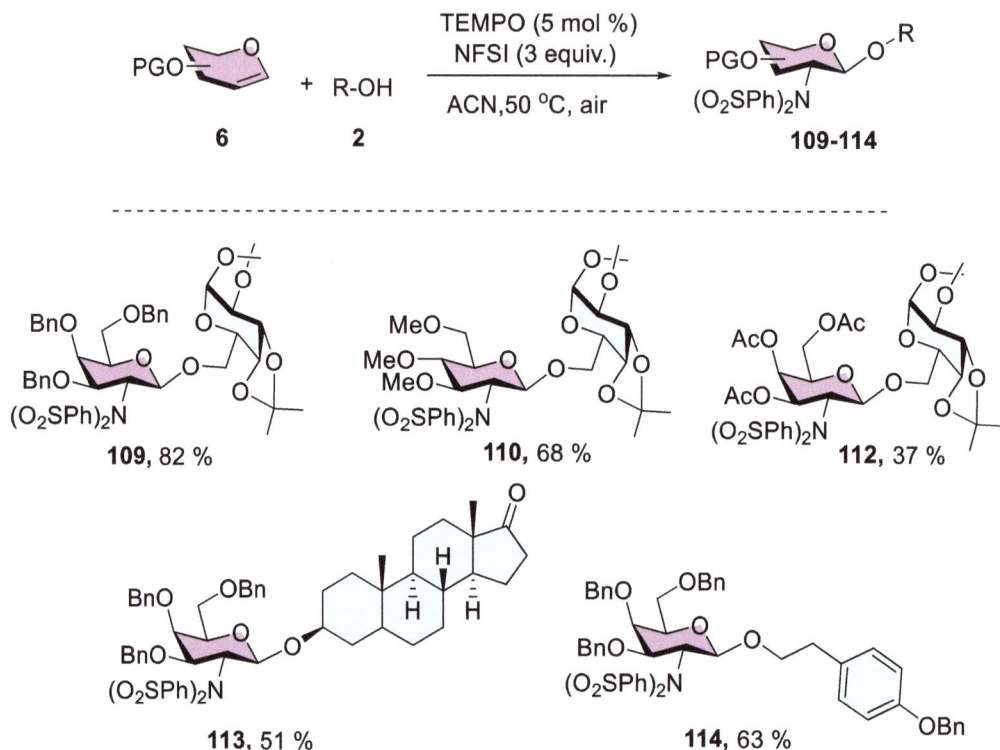

Scheme 26. Nitrogen-centered radical mediated cascade amidoglycosylation of glycals.

In 2021, P. Peng and co-workers [69] featured the platinum (IV) chloride-mediated glycosylation of *O*-glycosyl trichloro acetimidates **66**, facilitating the formation of β-glycosides in good yield, as shown in (Scheme **27**). The reaction was explored by different mono hydroxyl acceptors (**116-117**), giving the desired product with high yield and high β-selectivities. Secondary hydroxyl acceptors are less reactive due to steric hindrance; however, the reaction under optimized conditions proceeds well with excellent yield and selectivity (**118-120**). When *O*-mannopyranosyl trichloroacetimidate is used as the glycosyl acceptor, it undergoes the reaction to give the desired product in 75% yield with moderate β selectivity (**120**). The reaction with 2-azido-2-deoxy-3,4,6-tri-*O*-benzyl-α-D-glucopyranosyltrichloroacetimidate affords β-glycosides **121** in 85% yield.

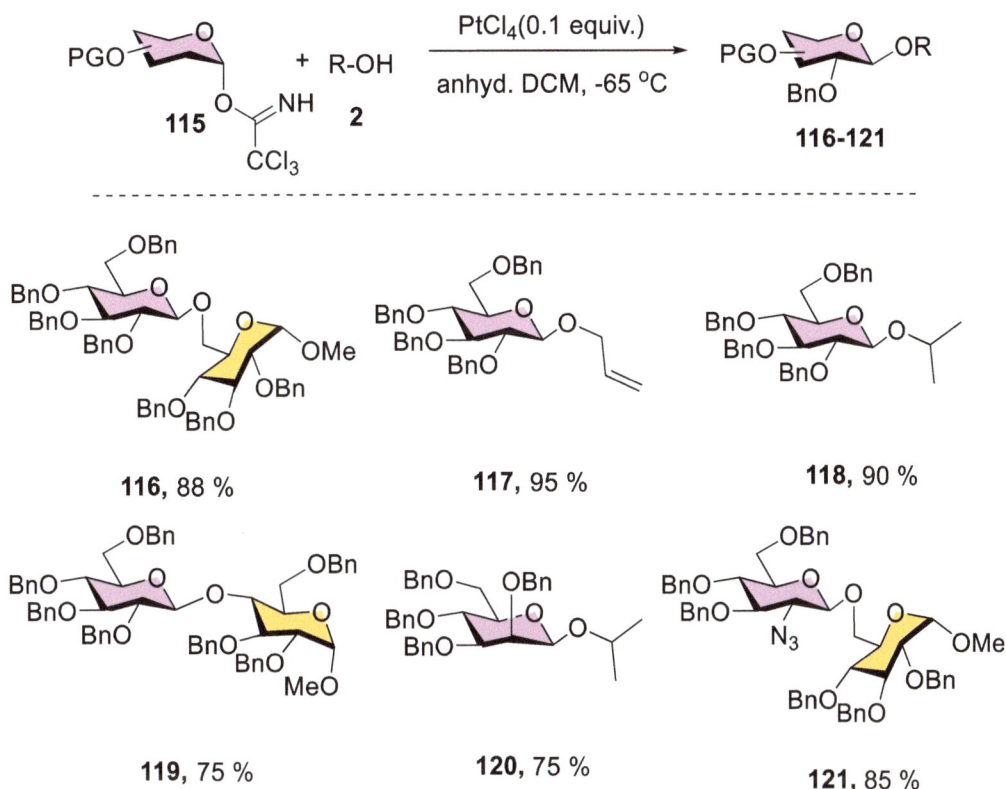

Scheme 27. *O*-glycosyl trichloroacetimidates as glycosyl donors and platinum (IV) chloride as a dual catalyst permitting stereo- and regioselective glycosidations.

In 2022, H. Yao and coworkers [70] discussed an open-air stereoselective *O*-glycosylation of glycals with aryl boronic acid, resulting in β-*O*-glycosides in good yield, with no *C*-glycosides observed (Scheme **28**). The substrate scope with 3,4-*O*-carbonate-L-arabinal was explored with different aryl boronic acids, such as *p*-methoxy **123** and 4-isopropoxyphenyl **124**. *m*-bromo substituents **125** gives the desired β *O*-glycosides in high yield. After exploring the broad scope of boronic acid, different glycals were also reacted, including 6-*O*-TBDPS-3,4-*O*-carbonate-D-galactal **126** and 6-*t*-butyloxycarbonyl-D-galactal **127**, yielding the desired product in good yield. Fucal donor **128** reacts with 3,4-dimethylphenyl boronic acids to give the β-L-fucoside in 81% yield. Isotope labeling and control experiments were also conducted in the research article to support the mechanism of the reaction.

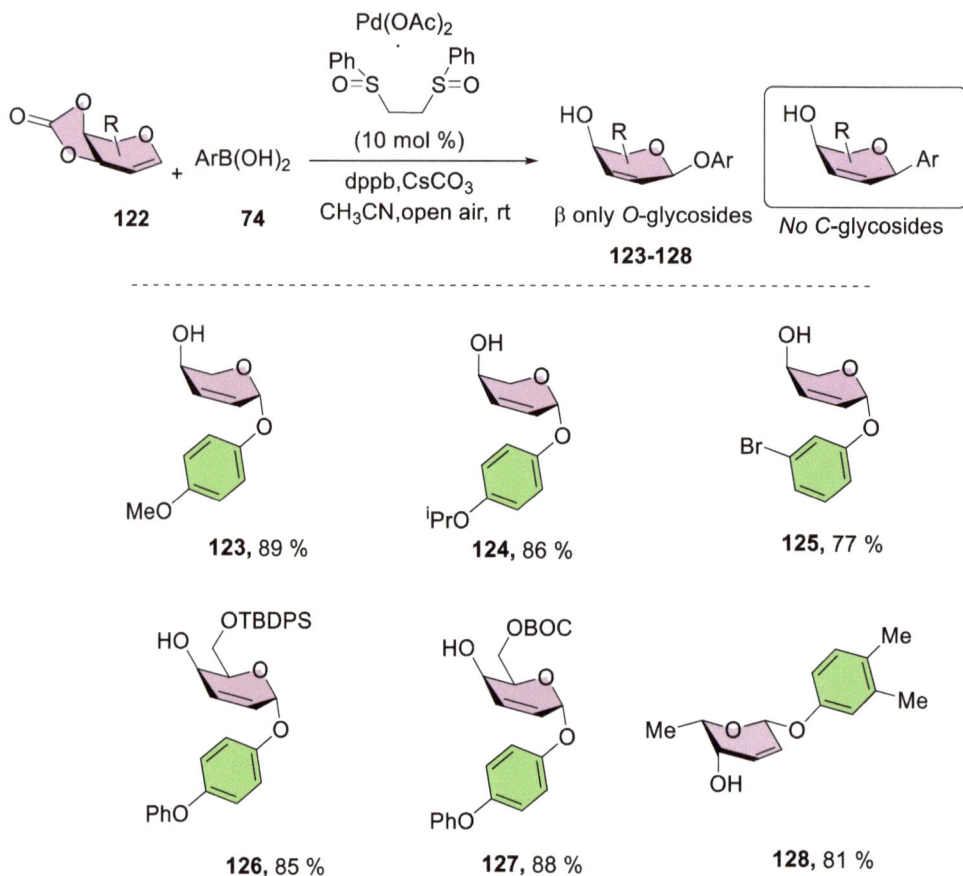

Scheme 28. Stereoselective *O*-glycosylation of glycals with arylboronic acids using air as the oxygen source.

In 2022, Kashyap and coworkers [71] disclosed the bismuth triflate catalyzed stereoselective glycosylation of armed and disarmed donors, as depicted in (Scheme **29**). This approach is based on solvent and temperature-controlled chemoselective activation of glycal donors to differentiate between the Ferrier and 2-deoxy products. They utilized disarmed glycals in 1,4-dioxane at 50 °C for further derivatization with different acceptors (primary and secondary alcohols) to give 2-deoxy galactosides **81**, **130-133**. Armed glycosyl donors give the desired product in toluene at room temperature **134**.

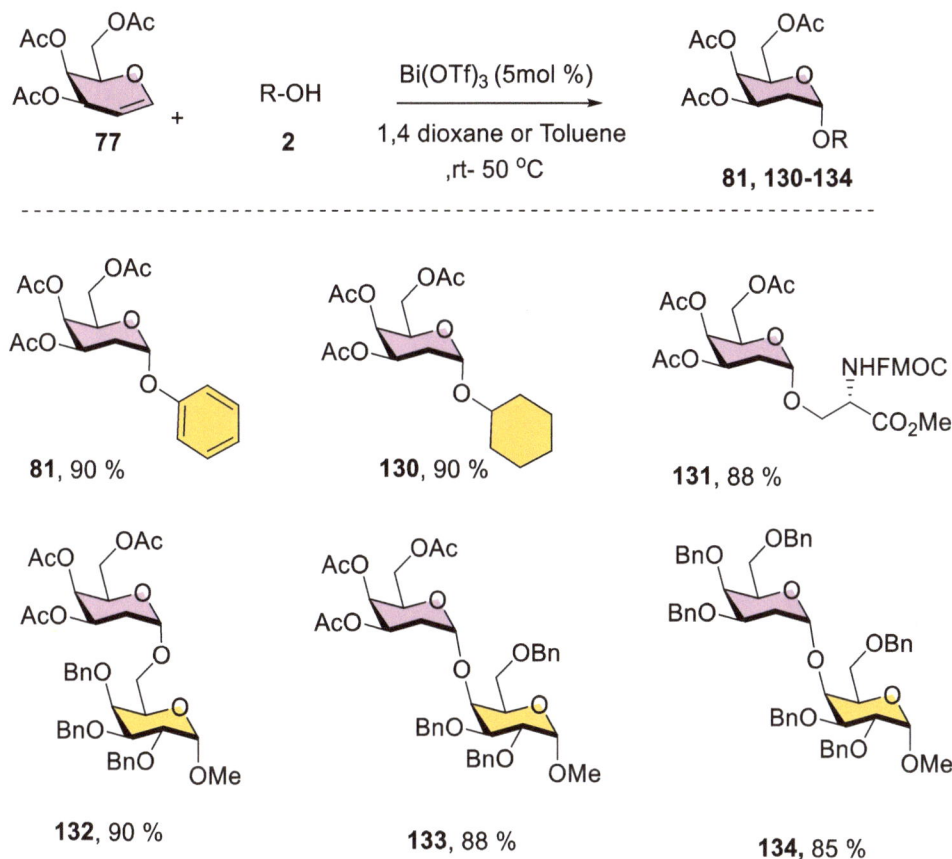

Scheme 29. Bismuth-catalyzed stereoselective 2-deoxyglycosylation of disarmed/armed glycal donors.

In 2023, H. Yao and coworkers [72] revealed the direct utilization of boron acceptors in the synthesis of *O*-glycosides, as shown in (Scheme **30**). Palladium catalyst-mediated reaction generates Ferrier β-glycosides, and a copper catalyst reaction gives 2-deoxy glycosides in good to excellent yield. Aromatic and aliphatic boron acceptors both give Ferrier products in high yield and with β-selectivity (**136-138**). Similarly, in the presence of copper triflate, the reaction proceeds to give 2-deoxy glycoside with α-selectivity and good yield (**139-141**).

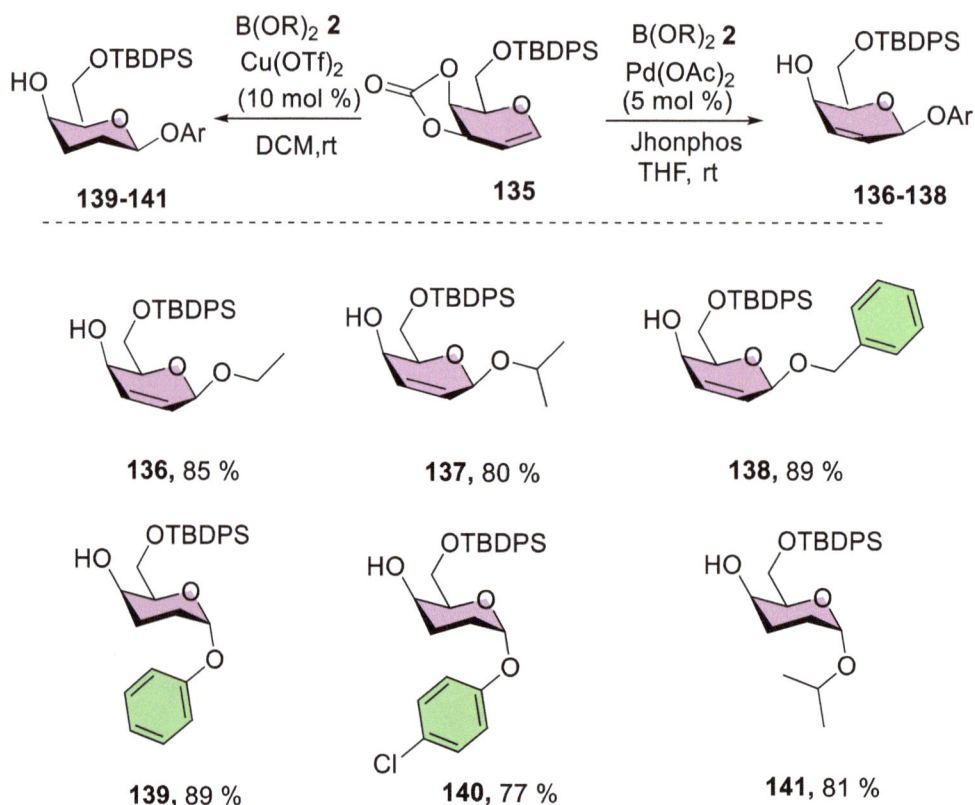

Scheme 30. Stereoselective synthesis of *O*-glycosides with borate acceptors.

In 2023, H. Yao and coworkers [73] developed the iron-catalyzed 2-deoxy glycosylation from 3,4-*O*-carbonate glycals **142** at room temperature (as shown in Scheme **31**). The substrate scope of 3,4-*O*-carbonate D-galactose with phenol (**143-144**) and alcohol derivatives (**145-146**) was explored, providing the desired product in good yield. The reaction was performed with different sugars **147**, which were also tested to yield glycoside in high yield. Natural products were also used as acceptors to give cholesterol α-2-deoxyglycoside **148**. This strategy was further extended to the sequential synthesis of oligosaccharides **104**, as shown in (Scheme **32**). Firstly, **142** gives the glycosylation reaction with **149**, resulting in the formation of 2-deoxydisaccharide **150**. **150** was deprotected by TBAF to afford **151** with a free C6-OH in 89% yield. Then, 2-deoxyglycosylation **151** with **142**, followed by deprotection, generated trisaccharide **152** in 56% yield in two steps. The glycosylation process was repeated once more, and finally, the tetrasaccharides **153** were generated in 47% yield with only α-stereoselectivity.

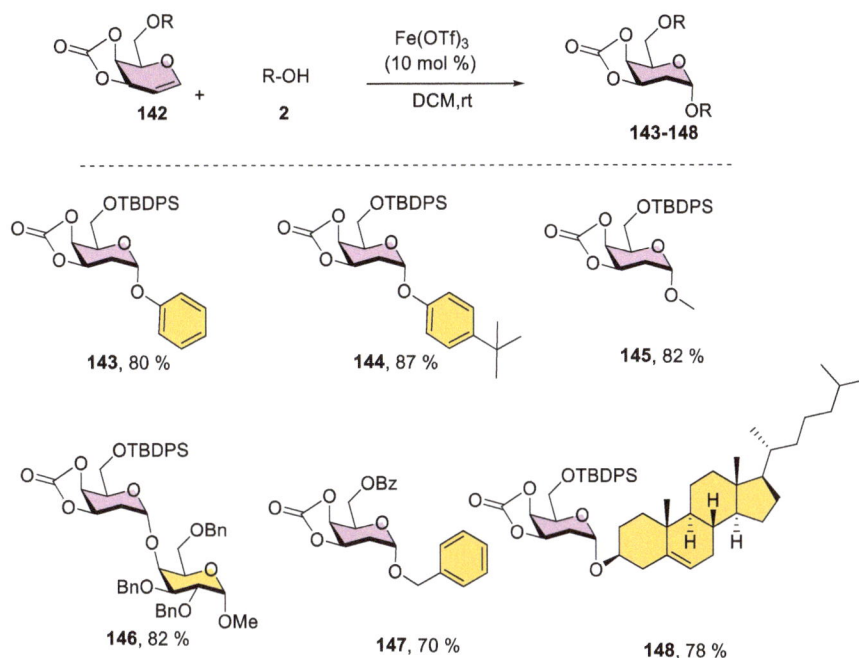

Scheme 31. Iron-catalyzed stereoselective synthesis of 2-deoxy glycosides.

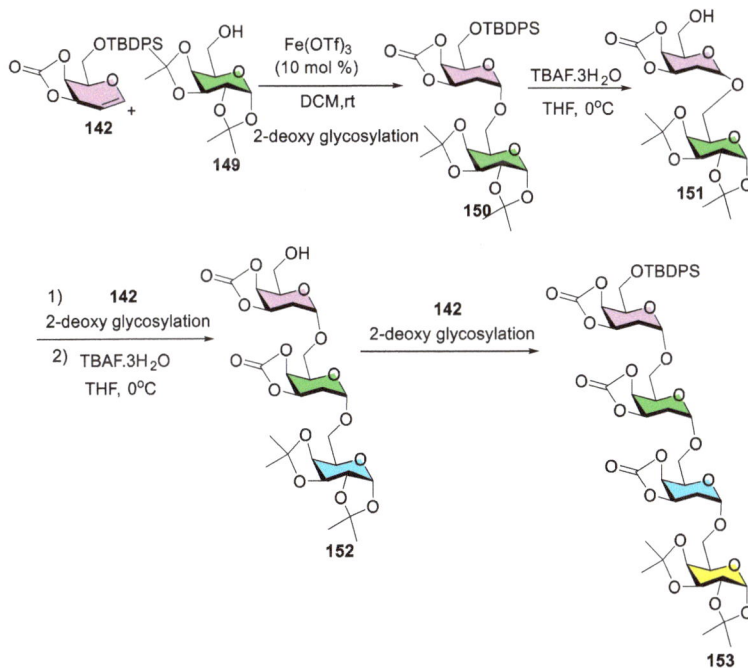

Scheme 32. Synthesis of tetrasaccharides.

In 2023, Mukherjee and coworkers [74] presented a one-pot reaction of glycals with two nucleophiles in the presence of tin chloride access to 2-deoxy-3-thio *O*-glycosides, as shown in (Scheme **33**). The reaction was performed with numerous glycals with different substituted thiophenols, giving the product in good to excellent yield (**155-157**). After that, glucal with thiophenol smoothly reacted with different acceptors to give 2-deoxy-3-thioaryl *O*-glycosides in one pot within 20 min in excellent yield (**158-160**).

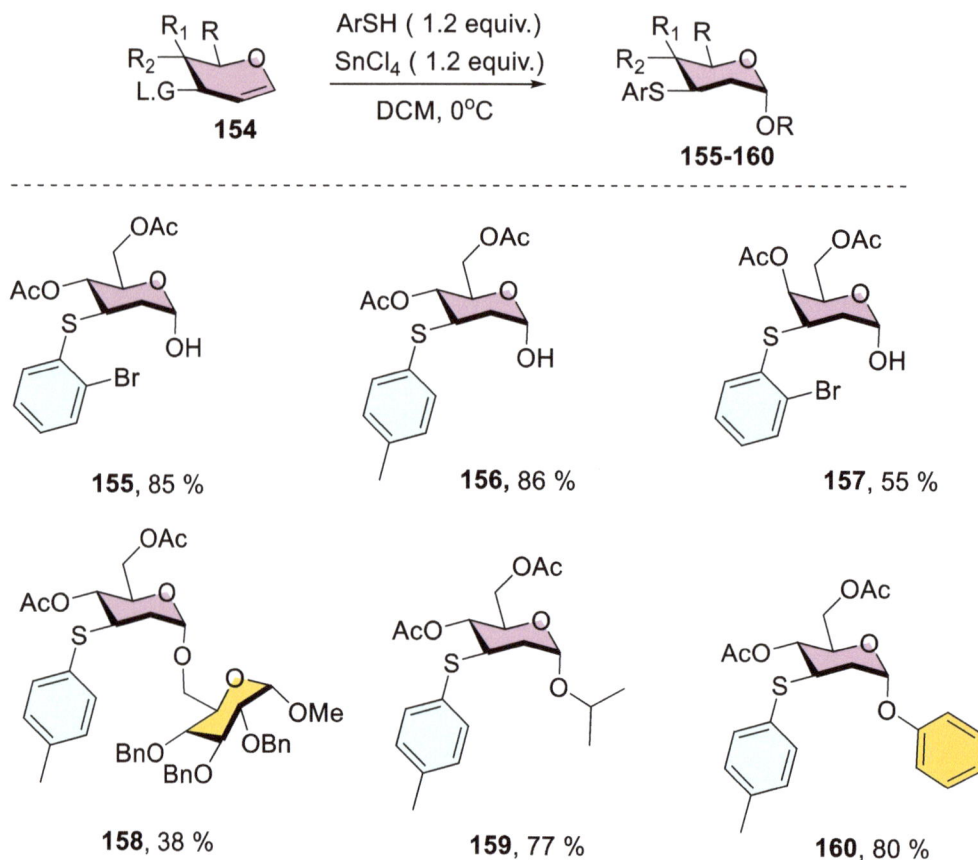

Scheme 33. Regio and stereoselective one-pot synthesis of 2-deoxy-3-thio pyranoses and their *O*-glycosides from glycal.

In 2024, C. Xu and coworkers [75] presented a noncovalent organocatalytic concerted addition of phenol to glycal to give stereoselective phenolic 2-deoxy glycoside, as shown in (Scheme **34**). Substrate scopes were explored by the glycosylation of glycals with different aglycons like naphthol, substituent phenols, *etc.*, yielding the desired product in good yield (**161-162**). Additionally, some bioactive phenolic molecules, including sesamol, estrone, *etc.*, under optimized reaction conditions, give the desired glycosides in good yield (**163-164**). Different sugar moieties were also tested, such as silylated arabinal and xylal, and were readily converted into corresponding glycosides in good yield (**165-166**). The reported methodology was based on anion-bridged dual hydrogen bond interaction, which was confirmed experimentally by NMR, UV-visible, and fluorescence analysis and further verified by DFT calculations.

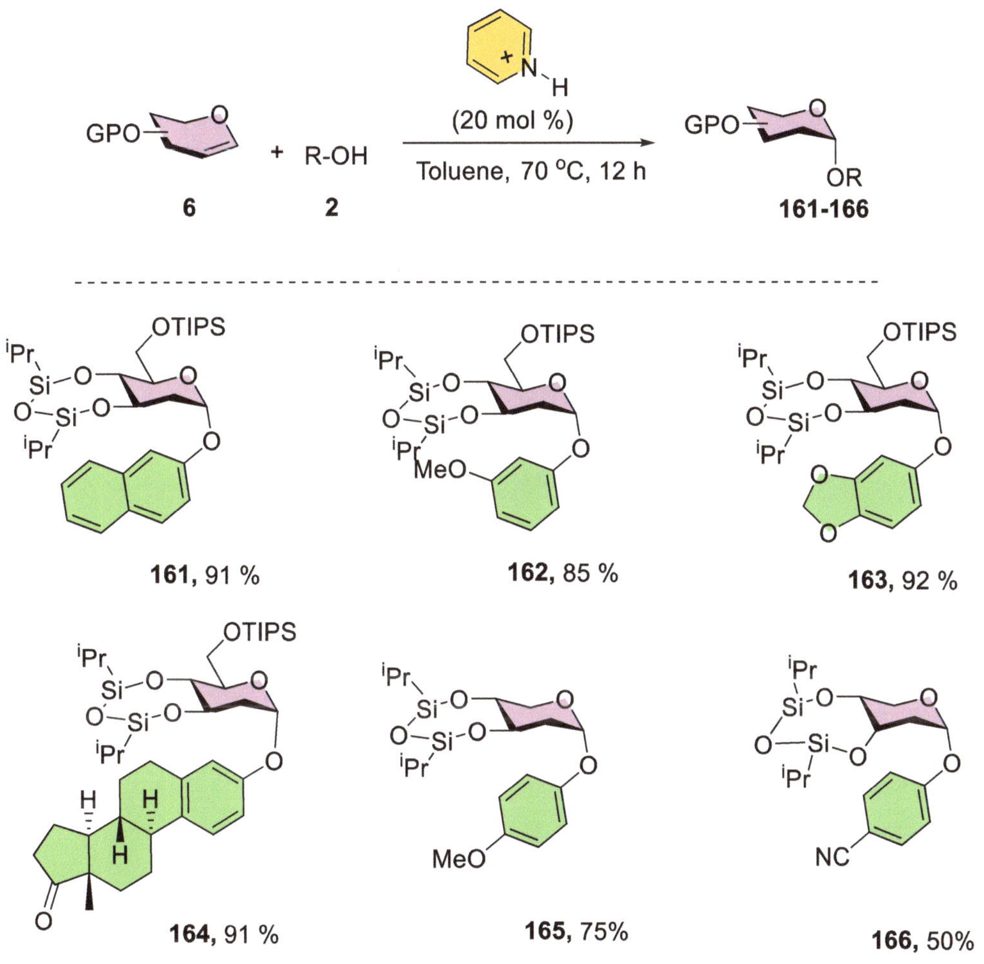

Scheme 34. Anion-bridged dual hydrogen bond enabled concerted addition of phenol to glycals.

Recently, Hussain and coworkers introduced an open-air approach for aldehyde-mediated selective β-glycosylation of glycals employing carboxylic acids as the nucleophile, as outlined in (Scheme **35**) [76]. The flexibility of this method was showcased by testing a range of substrates, leading to the successful formation of 2-hydroxy β-glycosyl ester derivatives with high yields. Various glycals, including acetyl-protected glucal, galactal, and silyl-protected glucal, delivered the corresponding products in good yields (**169-171**). In addition, substituted carboxylic acids were examined, including aliphatic acids, aromatic acids, and stearic acid, all of which yielded the desired products in good yields (**172-176**).

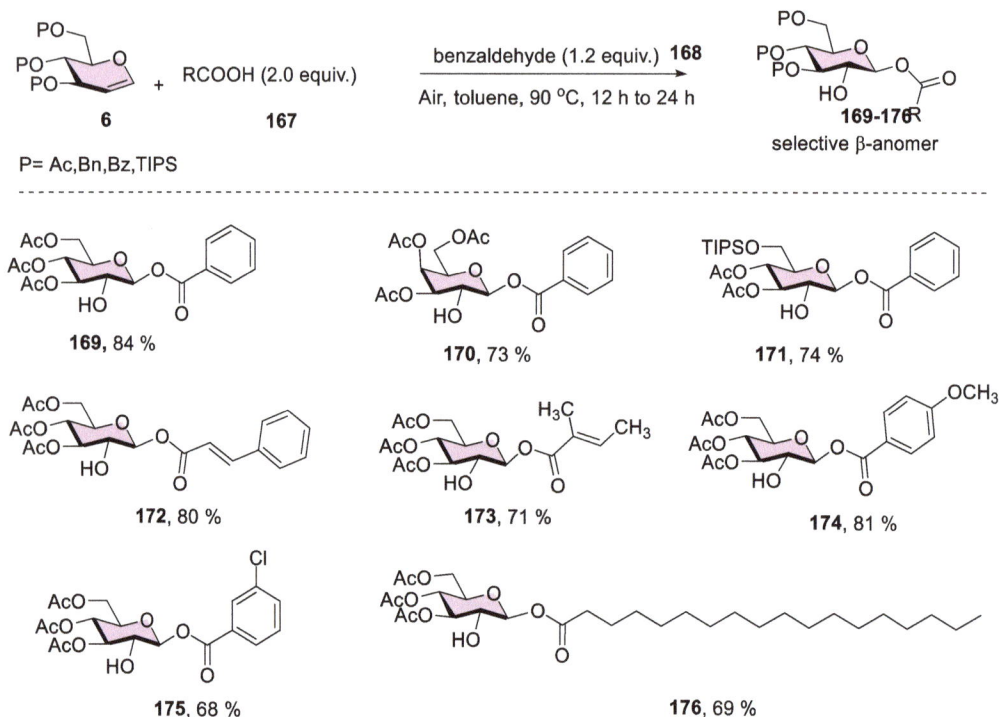

Scheme 35. Aldehyde-mediated open-air stereoselective β-glycosylation of glycals.

INTRODUCTION TO *S*-GLYCOSIDES

S-glycosides are a class of glycosides characterized by the presence of a sulfur atom in place of oxygen in the glycosidic bond between the sugar moiety and the aglycone [77]. These compounds are also referred to as thio-glycosides or glucosinolates. *S*-glycosides are naturally synthesized through the reaction between an NDP-sugar and a nucleophilic sulfur-containing compound. A notable example is the *S*-glycosylation of thiohydroamic acids with UDP-glucose, catalyzed by UDPG-thiohydroximate glucosyltransferase, resulting in the formation of desulfoglucosinolates, which are precursors to glucosinolates, as depicted in (Scheme **36**) [78].

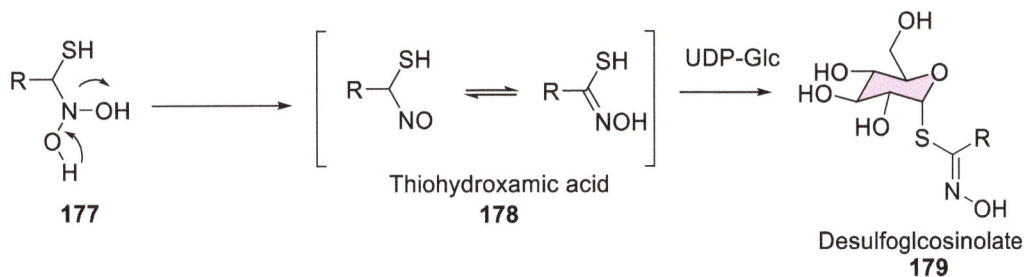

Scheme 36. *S*-glycosylation reaction for the biosynthesis of glucosinolates.

Medicinal Applications of S-glycosides

Numerous *S*-glycosides with intriguing biological activities have been isolated from a wide variety of organisms. Glucosinolates, which were introduced earlier, represent one of the most significant classes of compounds. Their distinctive properties were first noted in the 17th century, and to date, around 120 different glucosinolates have been discovered across 16 families of dicotyledonous angiosperms [79]. Glucosinolates are β-thioglucoside *N*-hydroxy sulfates, varying in their side-chain R (Scheme **37**), and serve as stable, water-soluble precursors to isothiocyanates. The cancer chemoprotective properties of glucosinolates and isothiocyanates are now garnering significant attention for the development of new anticancer drugs. Lincomycin is a thioglycoside-based antibiotic produced by *Streptomyces lincolnensis* and contains an aminooctose moiety. Lincomycin also exhibits bacteriostatic activity by inhibiting bacterial protein synthesis [80]. Sinigrin is a naturally occurring glucosinolate found in the seeds of black mustard. Sinigrin has demonstrated a range of beneficial properties including anti-cancer, antibacterial, antifungal, antioxidant, and anti-inflammatory effects, and is also known to promote wound healing and biofumigation [81].

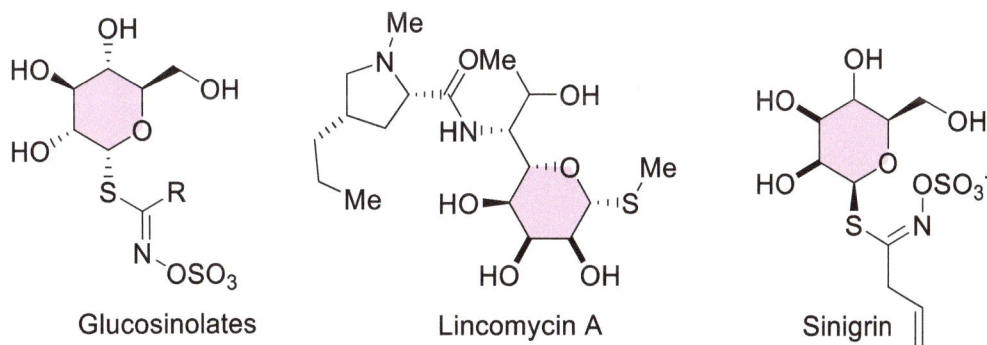

Scheme 37. Medicinal applications of *S*-glycosides.

Literature On Synthesizing S-glycosides

The synthesis of *S*-glycosides in the laboratory mirrors the natural biosynthesis processes, utilizing nucleophilic sulfur to displace a leaving group at the anomeric carbon of the sugar. Some recent synthetic strategies employing a detailed overview of the key aspects involved in the synthesis and stereochemical control of *S*-glycosides are discussed.

The research group of Jiang reported a stereoselective thioglycosylation of glycals **180** with Bunte salt **181** *via* Pd-catalyzed allylic rearrangement to access various thioglycosides **182** with α-selectivity, as depicted in (Scheme **38**) [82]. The optimal reaction conditions revealed the use of benzylthiosulfate salt as the sulfur source and $PdCl_2(PhCN)_2$/DPPF as the catalytic system in MeCN solvent for 12 h under a nitrogen atmosphere, resulting in products **182a-f** with good yields. The authors disclosed that a range of benzyl thioglycosides were selectively produced using different glycals derived from L-arabinose, L-rhamnose, and L-galactose.

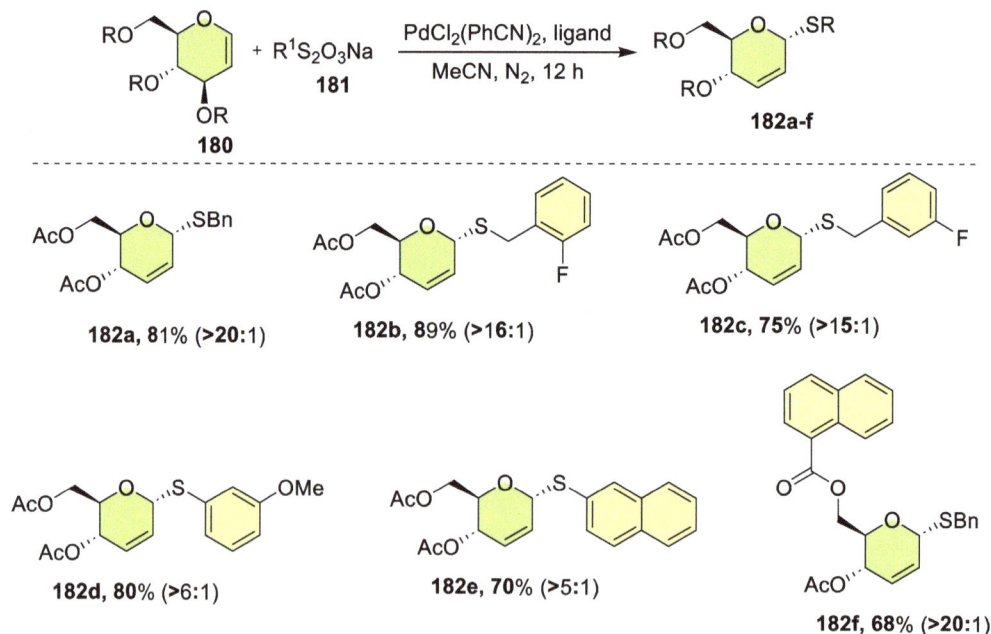

Scheme 38. Pd-catalyzed stereoselective thioglycosylation of glycals with Bunte salt.

A versatile *S*-glycosylation method was established by Yao *et al.* using Pd-catalyzed allylic substitution in the presence of xantphos ligand in DCM solvent to produce α- and β-thioglycosides, as illustrated in (Scheme **39**) [83]. A wide variety of substrates were effectively tolerated under the standard conditions, yielding diverse thioglycosides in a stereospecific manner with good efficiency. The authors noted that the stereochemistry of the resultant thiol moiety is influenced by the stereochemistry of the pre-attached group at the anomeric center.

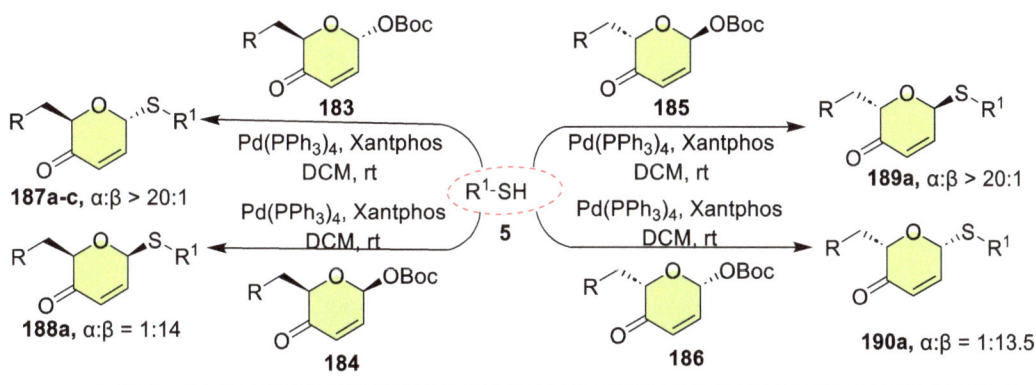

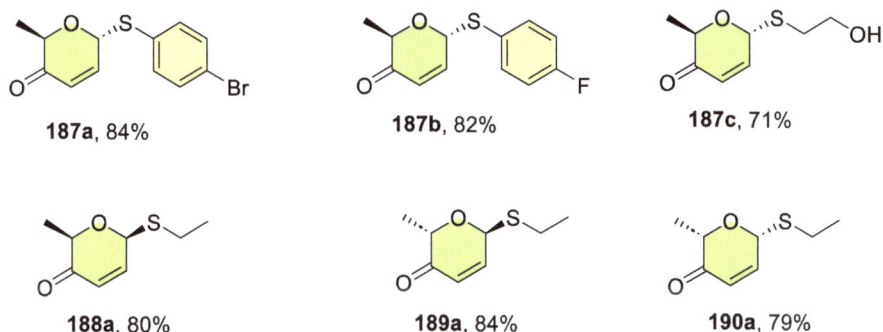

Scheme 39. Pd-catalyzed approach for the *S*-glycosylation.

The research group of Huang, Zou, and Yao reported an effective regiodivergent synthetic route for accessing 1- and 3-thiosugars catalyzed by palladium and cobalt catalysis, respectively, as illustrated in (Scheme **40**) [84]. The authors utilized 3,4-*O*-carbonate glycals **191** as donors and various aliphatic and aromatic thiols as acceptors. When Pd$_2$(dba)$_3$ was employed as the catalyst, only β-1-thiosugars **192** were produced. Conversely, the application of Co(BF$_4$)$_2$ catalyst led to the generation of 3*S*-3-thiosugars **193** with high regioselectivity and stereoselectivity.

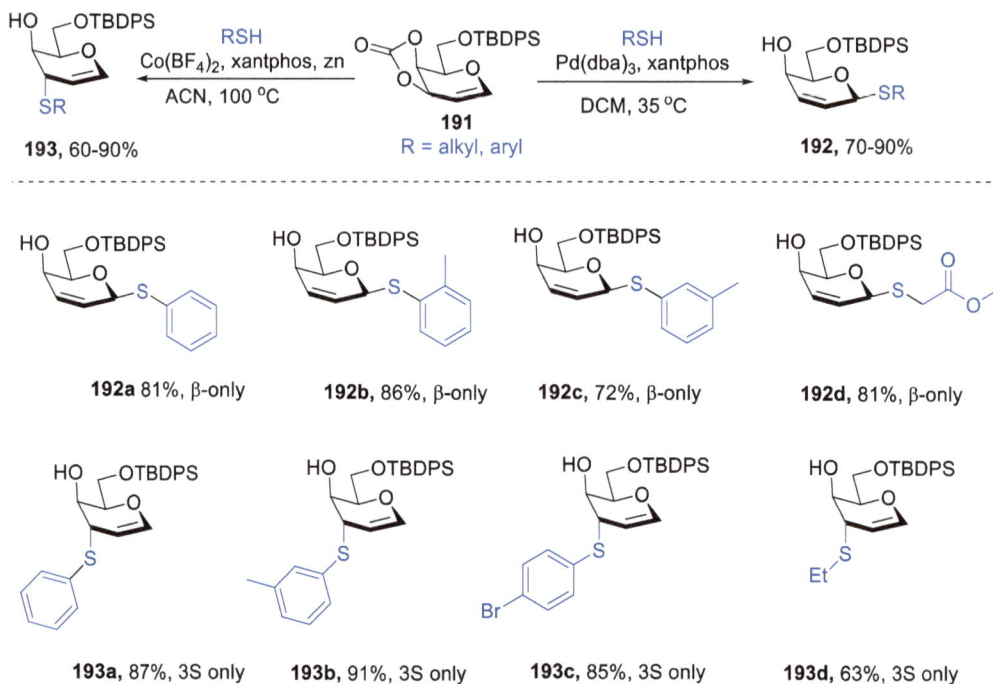

Scheme 40. Synthesis of 1- and 3-thiosugars catalyzed by palladium and cobalt, respectively.

Chen and their coworkers explored the combination of $Zn/AlCl_3$ and $Hf(OTf)_4$ complex to catalyze the reaction of glycals **194** and **195** with disulfide compounds R_2S_2 at 60 °C, yielding the desired thiosugars with good efficiency, as depicted in (Scheme **41**) [85]. In this system, disulfide served as the *S*-nucleophile, while $AlCl_3$ and $Hf(OTf)_4$ functioned as Lewis acids. The authors disclosed that both 1-thiosugars and 3-thiosugars could be produced, depending upon the attacking sites available on glycals.

CONCLUSION

The chemical *O*- and *S*-glycosylation strategies have shown noteworthy progress in recent times, reflecting their growing importance in the synthesis of biologically relevant scaffolds. The novel approaches devised in contemporary research provide enhanced insights into the reactivity of glycosyl donors and acceptors. The stereoselectivity of the desired products has been elucidated through the utilization of various promoters, solvent-controlled pathways, and temperature-controlled reactions. Several methodologies have been devised for the one-pot synthesis of oligosaccharides, reducing both labor and time consumption. However, challenges in oligosaccharide synthesis persist. There is a lack of environmentally friendly approaches in this domain, necessitating concerted efforts to develop atom-economic glycosylation and greener methodologies. Ultimately, this chapter offers a brief introduction to the recent strategies for chemical *O*- and *S*-glycosylation, aiming to aid researchers in comprehending the field and laying the groundwork for future advancements in this crucial area.

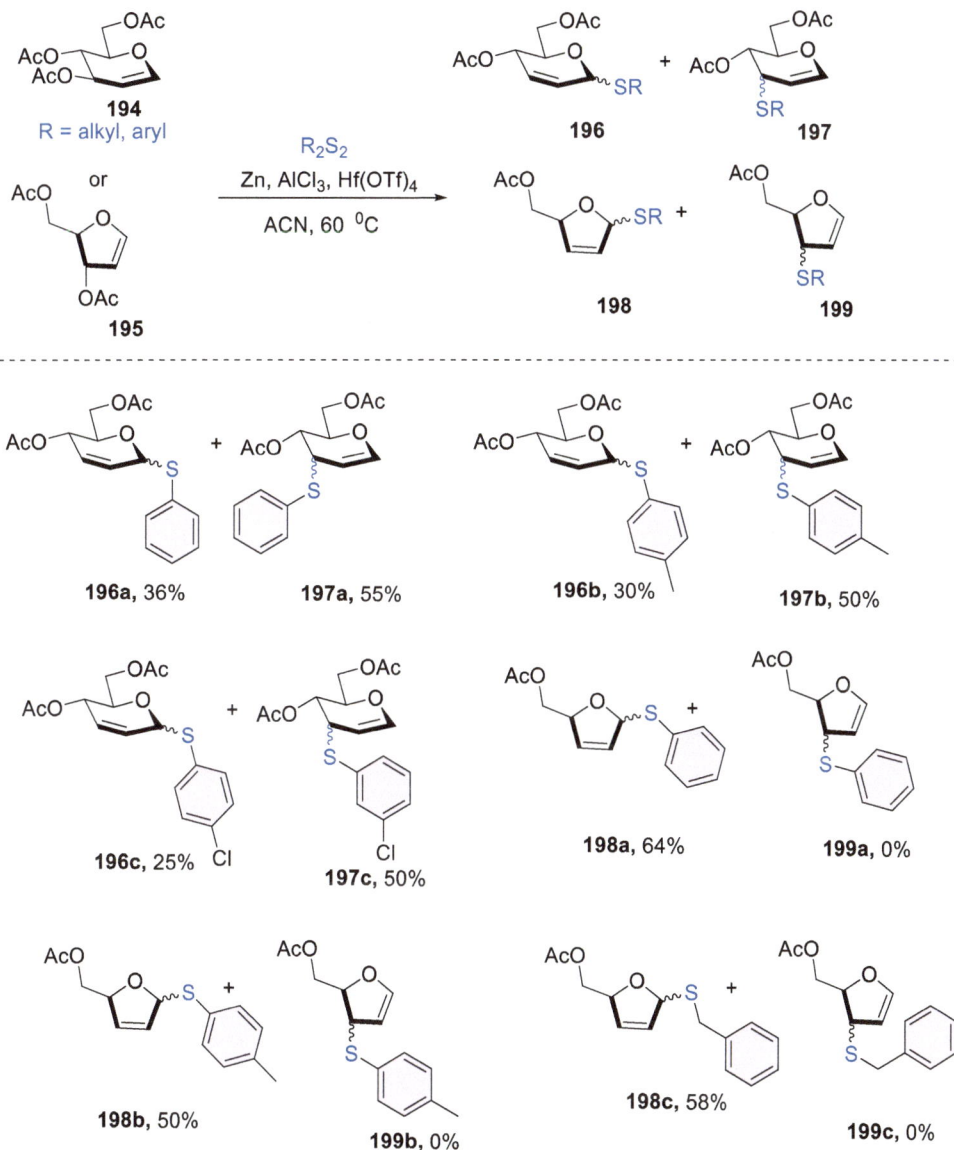

Scheme 41. Zn/AlCl$_3$ and Hf(OTf)$_4$-catalyzed thioglycosylation.

REFERENCES

[1]   Dondoni, A.; Marra, A. Methods for anomeric carbon-linked and fused sugar amino acid synthesis: the gateway to artificial glycopeptides. *Chem. Rev.,* **2000**, *100*(12), 4395-4422.
        [http://dx.doi.org/10.1021/cr9903003] [PMID: 11749352]

[2]   Lalitha, K.; Muthusamy, K.; Prasad, Y.S.; Vemula, P.K.; Nagarajan, S. Recent developments in β- C-glycosides: synthesis and applications. *Carbohydr. Res.,* **2015**, *402*, 158-171.
        [http://dx.doi.org/10.1016/j.carres.2014.10.008] [PMID: 25498016]

[3]   Montreuil, J. *Primary Structure of Glycoprotein Glycans Basis for the Molecular Biology of Glycoproteins*; **1980**, pp. 157-223.
        [http://dx.doi.org/10.1016/S0065-2318(08)60021-9]

[4]   Bartnik, M.; Facey, P.C. Glycosides. In: *Pharmacognosy*; Elsevier, **2017**; pp. 101-161.
        [http://dx.doi.org/10.1016/B978-0-12-802104-0.00008-1]

[5]   Alamgir, A. N. M. Secondary Metabolites: Secondary Metabolic Products Consisting of C and H; C, H, and O; N, S, and P Elements; and O/N Heterocycles; **2018**; pp 165–309.
        [http://dx.doi.org/10.1007/978-3-319-92387-1_3]

[6]   Soto-Blanco, B. Herbal Glycosides in Healthcare. In: *Herbal Biomolecules in Healthcare Applications*; Elsevier, **2022**; pp. 239-282.
        [http://dx.doi.org/10.1016/B978-0-323-85852-6.00021-4]

[7]   Bohé, L.; Crich, D. A propos of glycosyl cations and the mechanism of chemical glycosylation. *C. R. Chim.,* **2010**, *14*(1), 3-16.
        [http://dx.doi.org/10.1016/j.crci.2010.03.016]

[8]   Zhu, X.; Schmidt, R.R. New principles for glycoside-bond formation. *Angew. Chem. Int. Ed.,* **2009**, *48*(11), 1900-1934.
        [http://dx.doi.org/10.1002/anie.200802036] [PMID: 19173361]

[9]   Mulani, S.K.; Hung, W.C.; Ingle, A.B.; Shiau, K.S.; Mong, K.K.T. Modulating glycosylation with exogenous nucleophiles: an overview. *Org. Biomol. Chem.,* **2014**, *12*(8), 1184-1197.
        [http://dx.doi.org/10.1039/c3ob42129e] [PMID: 24382624]

[10]   Nicolaou, K.C.; Mitchell, H.J. Adventures in carbohydrate chemistry: new synthetic technologies, chemical synthesis, molecular design, and chemical biology a list of abbreviations can be found at the end of this article. telemachos charalambous was an inspiring teacher at the pancyprian gymnasium, nicosia, cyprus. *Angew. Chem. Int. Ed.,* **2001**, *40*(9), 1576-1624.
        [http://dx.doi.org/10.1002/1521-3773(20010504)40:9<1576::AID-ANIE15760>3.0.CO;2-G] [PMID: 11353467]

[11]   Koenigs, W.; Knorr, E. Ueber einige derivate des traubenzuckers und der galactose. *Ber. Dtsch. Chem. Ges.,* **1901**, *34*(1), 957-981.
        [http://dx.doi.org/10.1002/cber.190103401162]

[12]   Zimmermann, P.; Sommer, R.; Bär, T.; Schmidt, R.R. Azidosphingosine glycosylation in glycosphingolipid synthesis. *J. Carbohydr. Chem.,* **1988**, *7*(2), 435-452.
        [http://dx.doi.org/10.1080/07328308808058935]

[13]   Mukaiyama, T.; Murai, Y.; Shoda, S. An efficient method for glucosylation of hydroxy compounds using glucopyranosyl fluoride. *Chem. Lett.,* **1981**, *10*(3), 431-432.
        [http://dx.doi.org/10.1246/cl.1981.431]

[14]   Pozsgay, V.; Jennings, H.J. A new method for the synthesis of *O*-glycosides from *S*-glycosides. *J. Org. Chem.,* **1987**, *52*(20), 4635-4637.
        [http://dx.doi.org/10.1021/jo00229a047]

[15]   Jayakanthan, K.; Vankar, Y.D. Glycosyl trichloroacetylcarbamate: a new glycosyl donor for O-glycosylation. *Carbohydr. Res.,* **2005**, *340*(17), 2688-2692.
        [http://dx.doi.org/10.1016/j.carres.2005.07.024] [PMID: 16212950]

[16] Bettadaiah, B.K.; Srinivas, P. ZnCl2-catalyzed Ferrier reaction; synthesis of 2,3-unsaturated 1-O-glucopyranosides of allylic, benzylic and tertiary alcohols. *Tetrahedron Lett.,* **2003**, *44*(39), 7257-7259.
[http://dx.doi.org/10.1016/S0040-4039(03)01885-9]

[17] Fraser-Reid, B.; Radatus, B. 4,6-Di-O-acetyl-aldehydo-2,3-dideoxy-D-erythro-trans-hex-2-enose. Probable reason for the 'al' in Emil Fischer's triacetyl glucal. *J. Am. Chem. Soc.,* **1970**, *92*(17), 5288-5290.
[http://dx.doi.org/10.1021/ja00720a087]

[18] Hussain, N.; Hussain, A. Advances in Pd-catalyzed C–C bond formation in carbohydrates and their applications in the synthesis of natural products and medicinally relevant molecules. *RSC Advances,* **2021**, *11*(54), 34369-34391.
[http://dx.doi.org/10.1039/D1RA06351K] [PMID: 35497292]

[19] Kiely, D.E. Editorial. *J. Carbohydr. Chem.,* **1992**, *11*(1).
[http://dx.doi.org/10.1080/07328309208016137]

[20] Danishefsky, S.J.; Bilodeau, M.T. Glycals in organic synthesis: the evolution of comprehensive strategies for the assembly of oligosaccharides and glycoconjugates of biological consequence. *Angew. Chem. Int. Ed. Engl.,* **1996**, *35*(13-14), 1380-1419.
[http://dx.doi.org/10.1002/anie.199613801]

[21] Das, R.; Mukhopadhyay, B. Chemical O-Glycosylations: an overview. *ChemistryOpen,* **2016**, *5*(5), 401-433.
[http://dx.doi.org/10.1002/open.201600043] [PMID: 27777833]

[22] Mukherjee, M.M.; Ghosh, R.; Hanover, J.A. Recent advances in stereoselective chemical *O*-Glycosylation reactions. *Front. Mol. Biosci.,* **2022**, *9*, 896187.
[http://dx.doi.org/10.3389/fmolb.2022.896187] [PMID: 35775080]

[23] Bennett, C.S.; Galan, M.C. Methods for 2-Deoxyglycoside synthesis. *Chem. Rev.,* **2018**, *118*(17), 7931-7985.
[http://dx.doi.org/10.1021/acs.chemrev.7b00731] [PMID: 29953219]

[24] Seeberger, P.H.; Werz, D.B. Synthesis and medical applications of oligosaccharides. *Nature,* **2007**, *446*(7139), 1046-1051.
[http://dx.doi.org/10.1038/nature05819] [PMID: 17460666]

[25] Fischer, E. Ueber die glucoside der alkohole. *Ber. Dtsch. Chem. Ges.,* **1893**, *26*(3), 2400-2412.
[http://dx.doi.org/10.1002/cber.18930260327]

[26] Lemieux, R.U.; Hendriks, K.B.; Stick, R.V.; James, K. Halide ion catalyzed glycosidation reactions. Syntheses of. alpha.-linked disaccharides. *J. Am. Chem. Soc.,* **1975**, *97*(14), 4056-4062.
[http://dx.doi.org/10.1021/ja00847a032]

[27] Ness, R.K.; Fletcher, H.G., Jr Evidence that the supposed 3,5-Di-O-benzoyl-1,2-O-(1-hydroxybenzylidene)-α-D- ribose is Actually 1,3,5-Tri-O-benzoyl-α-D-ribose. *J. Am. Chem. Soc.,* **1956**, *78*(18), 4710-4714.
[http://dx.doi.org/10.1021/ja01599a048]

[28] Fraser-Reid, B.; Udodong, U.E.; Wu, Z.; Ottosson, H.; Merritt, J.R.; Rao, C.S.; Roberts, C.; Madsen, R. N-Pentenyl glycosides in organic chemistry: a contemporary example of serendipity. *Synlett,* **1992**, *1992*(12), 927-942.
[http://dx.doi.org/10.1055/s-1992-21543]

[29] Roy, R.; Andersson, F.O.; Letellier, M. "Active" and "latent" thioglycosyl donors in oligosaccharide synthesis. Application to the synthesis of α-sialosides. *Tetrahedron Lett.,* **1992**, *33*(41), 6053-6056.
[http://dx.doi.org/10.1016/S0040-4039(00)60004-7]

[30] Zhang, Z.; Ollmann, I.R.; Ye, X.S.; Wischnat, R.; Baasov, T.; Wong, C.H. Programmable One-Pot Oligosaccharide Synthesis. *J. Am. Chem. Soc.,* **1999**, *121*(4), 734-753.

[http://dx.doi.org/10.1021/ja982232s]

[31]   Douglas, N.L.; Ley, S.V.; Lücking, U.; Warriner, S.L. Tuning glycoside reactivity: New tool for efficient oligosaccharide synthesis. *J. Chem. Soc., Perkin Trans. 1,* **1998**, (1), 51-66.
       [http://dx.doi.org/10.1039/a705275h]

[32]   Boons, G.J.; Isles, S. Vinyl glycosides in oligosaccharide synthesis. 2. the use of allyl and vinyl glycosides in oligosaccharide synthesis. *J. Org. Chem.,* **1996**, *61*(13), 4262-4271.
       [http://dx.doi.org/10.1021/jo960131b] [PMID: 11667325]

[33]   Ehle, M.; Patel, C.; Giugliano, R. P. Digoxin: Clinical Highlights. *Crit Pathways Cardiol A J Evidence-Based Med,* **2011**, *10*(2), 93-98.
       [http://dx.doi.org/10.1097/HPC.0b013e318221e7dd]

[34]   Kumar, A.; De, T.; Mishra, A.; Mishra, A. Oleandrin: A cardiac glycosides with potent cytotoxicity. *Pharmacogn. Rev.,* **2013**, *7*(14), 131-139.
       [http://dx.doi.org/10.4103/0973-7847.120512] [PMID: 24347921]

[35]   Manunta, P.; Ferrandi, M.; Bianchi, G.; Hamlyn, J.M. Endogenous ouabain in cardiovascular function and disease. *J. Hypertens.,* **2009**, *27*(1), 9-18.
       [http://dx.doi.org/10.1097/HJH.0b013e32831cf2c6] [PMID: 19050443]

[36]   He, H.; Ding, W.D.; Bernan, V.S.; Richardson, A.D.; Ireland, C.M.; Greenstein, M.; Ellestad, G.A.; Carter, G.T. Lomaiviticins A and B, potent antitumor antibiotics from micromonospora lomaivitiensis. *J. Am. Chem. Soc.,* **2001**, *123*(22), 5362-5363.
       [http://dx.doi.org/10.1021/ja010129o] [PMID: 11457405]

[37]   Gould, J.C. Erythromycin in respiratory tract infection. *Scott. Med. J.,* **1977**, *22*(1_suppl) Suppl., 355-359.
       [http://dx.doi.org/10.1177/00369330770220S103] [PMID: 414355]

[38]   McGuire, J.M.; Boniece, W.S.; Higgens, C.E.; Hoehn, M.M.; Stark, W.M.; Westhead, J.; Wolfe, R.N. Tylosin, a New Antibiotic: I. Microbiological Studies. *Antibiot. Chemother. (Northfield Ill.),* **1961**, *11*(5), 320-327.

[39]   Satoi, S.; Muto, N.; Hayashi, M.; Fujii, T.; Otani, M. Mycinamicins, new macrolide antibiotics. I. Taxonomy, production, isolation, characterization and properties. *J. Antibiot. (Tokyo),* **1980**, *33*(4), 364-376.
       [http://dx.doi.org/10.7164/antibiotics.33.364] [PMID: 7410205]

[40]   Caltrider, P.G. Minomycin. In: *Mechanism of Action*; Springer Berlin Heidelberg: Berlin, Heidelberg, **1967**; pp. 669-670.
       [http://dx.doi.org/10.1007/978-3-642-46051-7_54]

[41]   Korynevska, A.; Heffeter, P.; Matselyukh, B.; Elbling, L.; Micksche, M.; Stoika, R.; Berger, W. Mechanisms underlying the anticancer activities of the angucycline landomycin E. *Biochem. Pharmacol.,* **2007**, *74*(12), 1713-1726.
       [http://dx.doi.org/10.1016/j.bcp.2007.08.026] [PMID: 17904109]

[42]   Honoré, N.; Cole, S.T. Streptomycin resistance in mycobacteria. *Antimicrob. Agents Chemother.,* **1994**, *38*(2), 238-242.
       [http://dx.doi.org/10.1128/AAC.38.2.238] [PMID: 8192450]

[43]   Nukada, T.; Bérces, A.; Whitfield, D.M. Can the stereochemical outcome of glycosylation reactions be controlled by the conformational preferences of the glycosyl donor? *Carbohydr. Res.,* **2002**, *337*(8), 765-774.
       [http://dx.doi.org/10.1016/S0008-6215(02)00043-5] [PMID: 11950473]

[44]   Kim, J.H.; Yang, H.; Park, J.; Boons, G.J. A general strategy for stereoselective glycosylations. *J. Am. Chem. Soc.,* **2005**, *127*(34), 12090-12097.
       [http://dx.doi.org/10.1021/ja052548h] [PMID: 16117550]

[45]   Smoot, J.T.; Pornsuriyasak, P.; Demchenko, A.V. Development of an arming participating group for

stereoselective glycosylation and chemoselective oligosaccharide synthesis. *Angew. Chem. Int. Ed.,* **2005**, *44*(43), 7123-7126.
[http://dx.doi.org/10.1002/anie.200502694] [PMID: 16224750]

[46]   Edward, J.T. *Anomeric Effect*; **1993**, pp. 1-5.
[http://dx.doi.org/10.1021/bk-1993-0539.ch001]

[47]   Romers, C.; Altona, C.; Buys, H. R.; Havinga, E. Geometry and Conformational Properties of Some Five- and Six-Membered Heterocyclic Compounds Containing Oxygen or Sulfur; **1969**; pp. 39–97.
[http://dx.doi.org/10.1002/9780470147139.ch2]

[48]   Satoh, H.; Hansen, H.S.; Manabe, S.; van Gunsteren, W.F.; Hünenberger, P.H. Theoretical investigation of solvent effects on glycosylation reactions: stereoselectivity controlled by preferential conformations of the intermediate oxacarbenium-counterion complex. *J. Chem. Theory Comput.,* **2010**, *6*(6), 1783-1797.
[http://dx.doi.org/10.1021/ct1001347] [PMID: 26615839]

[49]   Mong, K.T.; Nokami, T.; Tran, N.T.T.; Nhi, P.B. Solvent Effect on Glycosylation. In: *Selective Glycosylations: Synthetic Methods and Catalysts*; Wiley, **2017**; pp. 59-77.
[http://dx.doi.org/10.1002/9783527696239.ch3]

[50]   Balmond, E.I.; Benito-Alifonso, D.; Coe, D.M.; Alder, R.W.; McGarrigle, E.M.; Galan, M.C. A 3,4-trans-fused cyclic protecting group facilitates α-selective catalytic synthesis of 2-deoxyglycosides. *Angew. Chem. Int. Ed.,* **2014**, *53*(31), 8190-8194.
[http://dx.doi.org/10.1002/anie.201403543] [PMID: 24953049]

[51]   Murakami, T.; Sato, Y.; Yoshioka, K.; Tanaka, M. Novel stereocontrolled amidoglycosylation of alcohols with acetylated glycals and sulfamate ester. *RSC Advances,* **2014**, *4*(41), 21584-21587.
[http://dx.doi.org/10.1039/C4RA02367F]

[52]   Roy, R.; Rajasekaran, P.; Mallick, A.; Vankar, Y.D. Gold(III) Chloride and Phenylacetylene: A Catalyst System for the Ferrier Rearrangement, and *O* -Glycosylation of 1- *O* -Acetyl Sugars as Glycosyl Donors. *Eur. J. Org. Chem.,* **2014**, *2014*(25), 5564-5573.
[http://dx.doi.org/10.1002/ejoc.201402606]

[53]   Chen, P.; Su, J. Gd(OTf) 3 catalyzed preparation of 2,3-unsaturated O -, S -, N -, and C -pyranosides from glycals by Ferrier Rearrangement. *Tetrahedron,* **2016**, *72*(1), 84-94.
[http://dx.doi.org/10.1016/j.tet.2015.11.002]

[54]   Thombal, R.S.; Jadhav, V.H. Facile O-glycosylation of glycals using Glu-Fe 3 O 4 -SO 3 H, a magnetic solid acid catalyst. *RSC Advances,* **2016**, *6*(37), 30846-30851.
[http://dx.doi.org/10.1039/C6RA03305A]

[55]   Sau, A.; Galan, M.C. Palladium-Catalyzed α-Stereoselective *O* -Glycosylation of O(3)-Acylated Glycals. *Org. Lett.,* **2017**, *19*(11), 2857-2860.
[http://dx.doi.org/10.1021/acs.orglett.7b01092] [PMID: 28514163]

[56]   Palo-Nieto, C.; Sau, A.; Galan, M.C. Gold(I)-Catalyzed Direct Stereoselective Synthesis of Deoxyglycosides from Glycals. *J. Am. Chem. Soc.,* **2017**, *139*(40), 14041-14044.
[http://dx.doi.org/10.1021/jacs.7b08898] [PMID: 28934850]

[57]   Yao, H.; Zhang, S.; Leng, W.L.; Leow, M.L.; Xiang, S.; He, J.; Liao, H.; Le Mai Hoang, K.; Liu, X.W. Catalyst-Controlled Stereoselective *O* -Glycosylation: Pd(0) *vs.* Pd(II). *ACS Catal.,* **2017**, *7*(8), 5456-5460.
[http://dx.doi.org/10.1021/acscatal.7b01630]

[58]   Kabotso, D.E.K.; Pohl, N.L.B. Pentavalent Bismuth as a Universal Promoter for S-Containing Glycosyl Donors with a Thiol Additive. *Org. Lett.,* **2017**, *19*(17), 4516-4519.
[http://dx.doi.org/10.1021/acs.orglett.7b02080] [PMID: 28809575]

[59]   Tanaka, M.; Nakagawa, A.; Nishi, N.; Iijima, K.; Sawa, R.; Takahashi, D.; Toshima, K. Boronic-acid-catalyzed regioselective and 1,2- *cis* -stereoselective glycosylation of unprotected sugar acceptors *via*

SNi-Type Mechanism. *J. Am. Chem. Soc.,* **2018**, *140*(10), 3644-3651.
[http://dx.doi.org/10.1021/jacs.7b12108] [PMID: 29457892]

[60]   Yang, T.; Zhu, F.; Walczak, M.A. Stereoselective oxidative glycosylation of anomeric nucleophiles with alcohols and carboxylic acids. *Nat. Commun.,* **2018**, *9*(1), 3650.
[http://dx.doi.org/10.1038/s41467-018-06016-4] [PMID: 30194299]

[61]   Tatina, M.B.; Moussa, Z.; Xia, M.; Judeh, Z.M.A. Perfluorophenylboronic acid-catalyzed direct α-stereoselective synthesis of 2-deoxygalactosides from deactivated peracetylated d-galactal. *Chem. Commun. (Camb.),* **2019**, *55*(81), 12204-12207.
[http://dx.doi.org/10.1039/C9CC06151G] [PMID: 31549691]

[62]   Wang, J.; Deng, C.; Zhang, Q.; Chai, Y. Tuning the chemoselectivity of silyl protected rhamnals by temperature and brønsted acidity: kinetically controlled 1,2-addition *vs.* thermodynamically controlled ferrier rearrangement. *Org. Lett.,* **2019**, *21*(4), 1103-1107.
[http://dx.doi.org/10.1021/acs.orglett.9b00009] [PMID: 30714737]

[63]   Pal, K.B.; Guo, A.; Das, M.; Báti, G.; Liu, X.W. Superbase-catalyzed stereo- and regioselective glycosylation with 2-nitroglycals: facile access to 2-amino-2-deoxy- *O*-glycosides. *ACS Catal.,* **2020**, *10*(12), 6707-6715.
[http://dx.doi.org/10.1021/acscatal.0c00753]

[64]   Palo-Nieto, C.; Sau, A.; Jeanneret, R.; Payard, P.A.; Salamé, A.; Martins-Teixeira, M.B.; Carvalho, I.; Grimaud, L.; Galan, M.C. Copper reactivity can be tuned to catalyze the stereoselective synthesis of 2-deoxyglycosides from glycals. *Org. Lett.,* **2020**, *22*(5), 1991-1996.
[http://dx.doi.org/10.1021/acs.orglett.9b04525] [PMID: 32073274]

[65]   Kumar, M.; Reddy, T.R.; Gurawa, A.; Kashyap, S. Copper (ii)-catalyzed stereoselective 1,2-addition *vs.* Ferrier glycosylation of "armed" and "disarmed" glycal donors. *Org. Biomol. Chem.,* **2020**, *18*(25), 4848-4862.
[http://dx.doi.org/10.1039/D0OB01042A] [PMID: 32608448]

[66]   Zhang, J.; Dong, Y.; Yuma, M.; Mei, Y.; Jiang, N.; Yang, G.; Wang, Z. Copper-catalyzed stereoselective synthesis of 2-deoxygalactosides. *Synlett,* **2020**, *31*(11), 1087-1093.
[http://dx.doi.org/10.1055/s-0040-1707098]

[67]   Pal, K.B.; Guo, A.; Das, M.; Lee, J.; Báti, G.; Yip, B.R.P.; Loh, T-P.; Liu, X-W. Iridium-promoted deoxyglycoside synthesis: stereoselectivity and mechanistic insight. *Chem. Sci. (Camb.),* **2021**, *12*(6), 2209-2216.
[http://dx.doi.org/10.1039/D0SC06529C]

[68]   Shang, W.; Zhu, C.; Peng, F.; Pan, Z.; Ding, Y.; Xia, C. Nitrogen-centered radical-mediated cascade amidoglycosylation of glycals. *Org. Lett.,* **2021**, *23*(4), 1222-1227.
[http://dx.doi.org/10.1021/acs.orglett.0c04178] [PMID: 33560134]

[69]   Wan, Y.; Wu, X.; Xue, Y.; Lin, X-E.; Wang, L.; Sun, J-S.; Zhang, Q. Stereoselective glycosylation with conformation-constrained 2-Nitroglycals as donors and bifunctional thiourea as catalyst. *J. Carbohydr. Chem.,* **2021**, *40*(7-9), 535-557.
[http://dx.doi.org/10.1080/07328303.2021.2023560]

[70]   Wang, Q.; Lai, M.; Luo, H.; Ren, K.; Wang, J.; Huang, N.; Deng, Z.; Zou, K.; Yao, H. Stereoselective o-glycosylation of glycals with arylboronic acids using air as the oxygen source. *Org. Lett.,* **2022**, *24*(8), 1587-1592.
[http://dx.doi.org/10.1021/acs.orglett.1c04378] [PMID: 35080399]

[71]   Kumar, M.; Gurawa, A.; Kumar, N.; Kashyap, S. Bismuth-catalyzed stereoselective 2-deoxyglycosylation of disarmed/armed glycal donors. *Org. Lett.,* **2022**, *24*(2), 575-580.
[http://dx.doi.org/10.1021/acs.orglett.1c04008] [PMID: 34995079]

[72]   Zhao, X.; Zhang, Z.; Xu, J.; Wang, N.; Huang, N.; Yao, H. Stereoselective synthesis of *O* -Glycosides with borate acceptors. *J. Org. Chem.,* **2023**, *88*(16), 11735-11747.
[http://dx.doi.org/10.1021/acs.joc.3c01011] [PMID: 37525574]

[73]   Hou, M.; Xiang, Y.; Gao, J.; Zhang, J.; Wang, N.; Shi, H.; Huang, N.; Yao, H. Stereoselective synthesis of 2-deoxy glycosides *via* iron catalysis. *Org. Lett.,* **2023**, *25*(5), 832-837.
[http://dx.doi.org/10.1021/acs.orglett.2c04379] [PMID: 36700622]

[74]   Bhardwaj, M.; Mukherjee, D. Regio and stereoselective one-pot synthesis of 2-deoxy-3-thio pyranoses and their *o* -glycosides from glycals. *J. Org. Chem.,* **2023**, *88*(9), 5676-5686.
[http://dx.doi.org/10.1021/acs.joc.3c00146] [PMID: 37083468]

[75]   Jiao, Q.; Guo, Z.; Zheng, M.; Lin, W.; Liao, Y.; Yan, W.; Liu, T.; Xu, C. Anion-bridged dual hydrogen bond enabled concerted addition of phenol to glycal. *Adv. Sci. (Weinh.),* **2024**, *11*(11), 2308513.
[http://dx.doi.org/10.1002/advs.202308513] [PMID: 38225720]

[76]   Maheshwari, M.; Hussain, N. Aldehyde-mediated open-air stereoselective β-glycosylation of glycals: an expeditious route towards glycosyl-ester synthesis. *Adv. Synth. Catal.,* **2024**, *366*(21), 4478-4484.
[http://dx.doi.org/10.1002/adsc.202400638]

[77]   Romanò, C.; Jiang, H.; Boos, I.; Clausen, M.H. *S* -Glycosides: synthesis of *S* -linked arabinoxylan oligosaccharides. *Org. Biomol. Chem.,* **2020**, *18*(14), 2696-2701.
[http://dx.doi.org/10.1039/D0OB00470G] [PMID: 32206767]

[78]   Sørensen, H. Glucosinolates: structure-properties-function. In: *Canola and Rapeseed*; Springer US: Boston, MA, **1990**; pp. 149-172.
[http://dx.doi.org/10.1007/978-1-4615-3912-4_9]

[79]   Fahey, J.W.; Zalcmann, A.T.; Talalay, P. The chemical diversity and distribution of glucosinolates and isothiocyanates among plants. *Phytochemistry,* **2001**, *56*(1), 5-51.
[http://dx.doi.org/10.1016/S0031-9422(00)00316-2] [PMID: 11198818]

[80]   Abdul-Jabbar, A.M.; Hussian, N.N.; Mohammed, H.A.; Aljarbou, A.; Akhtar, N.; Khan, R.A. Combined anti-bacterial actions of lincomycin and freshly prepared silver nanoparticles: overcoming the resistance to antibiotics and enhancement of the bioactivity. *Antibiotics (Basel),* **2022**, *11*(12), 1791.
[http://dx.doi.org/10.3390/antibiotics11121791] [PMID: 36551448]

[81]   Mazumder, A.; Dwivedi, A.; Du Plessis, J. Sinigrin and its therapeutic benefits. *Molecules,* **2016**, *21*(4), 416.
[http://dx.doi.org/10.3390/molecules21040416] [PMID: 27043505]

[82]   Li, J.; Wang, M.; Jiang, X. Diastereoselective synthesis of thioglycosides *via* pd-catalyzed allylic rearrangement. *Org. Lett.,* **2021**, *23*(23), 9053-9057.
[http://dx.doi.org/10.1021/acs.orglett.1c03302] [PMID: 34783571]

[83]   Wang, Y.; Cao, Z.; Wang, N.; Liu, M.; Zhou, H.; Wang, L.; Huang, N.; Yao, H. Palladium-catalyzed stereospecific *s* -glycosylation by allylic substitution. *Adv. Synth. Catal.,* **2023**, *365*(10), 1699-1704.
[http://dx.doi.org/10.1002/adsc.202300129]

[84]   Liu, Y.; Jiao, Y.; Luo, H.; Huang, N.; Lai, M.; Zou, K.; Yao, H. Catalyst-controlled regiodivergent synthesis of 1- and 3-thiosugars with high stereoselectivity and chemoselectivity. *ACS Catal.,* **2021**, *11*(9), 5287-5293.
[http://dx.doi.org/10.1021/acscatal.1c00225]

[85]   Chen, P.; Guo, S.; Zuo, J.; Chu, R.; He, X.; Zhu, G. Synthesis of 3-S- and 3-Se-glycals by using R---S- R and R-Se-Se-R as the nucleophile precursors promoted by Hf(OTf)$_4$ and the temperature-dependent formation of the above-mentioned 3-S- and 3-Se products. *Tetrahedron Lett.,* **2020**, *61*(12), 151648.
[http://dx.doi.org/10.1016/j.tetlet.2020.151648] [PMID: 32153306]

CHAPTER 3

Recent Advances in the Synthesis of *C*-Glycosides Using Glycals and their Derivatives

Manish Kumar Sharma[1], Anand Kumar Pandey[1], Ram Pratap Pandey[1] and Nazar Hussain[1,*]

[1] *Department of Medicinal Chemistry, Institute of Medical Sciences, Banaras Hindu University, Varanasi-221005, India*

Abstract: A glycoside is recognized as a *C*-glycoside when the anomeric carbon of a sugar moiety is linked to an aglycone or another carbohydrate, creating a new C-C bond. Recently, *C*-glycosides have been recognized as a privileged class of carbohydrates owing to their extensive application in medicinal chemistry and drug discovery. These glycosides have played a vital role in designing various drug candidates owing to their stability towards acidic or enzymatic degradation. This chapter deals with the recent advances in the synthesis of privileged *C*-glycosides, including 2-deoxy-*C*-glycosides, using glycals and their derivatives.

Keywords: *C*-glycosides, Cross-coupling reactions, Directing groups, Drug discovery, Ferrier rearrangement, Glycals, Glycoconjugates, Glycosylation, Natural products, Photoredox catalysis, Transition metal catalysis.

C-glycosides are an important class of carbohydrate derivatives where a carbohydrate unit is directly connected to an aglycone or another carbohydrate, creating a new C-C bond [1 - 4]. Recently, *C*-glycosides have been recognized as an important carbohydrate mimetic due to their extensive application in medicinal chemistry and drug discovery programs [5]. They are known to exhibit high chemical and metabolic stability as compared to *O*-glycosides. The recent advancements in the synthesis of *C*-glycosides have reignited the interest of synthetic chemists by substituting the enzymatically labile glycosidic oxygen with a carbon center. Furthermore, the stability of *C*-glycosides toward acidic or enzymatic degradation has played a pivotal role in the designing of biological probes and pharmaceutical compounds. For example, a number of SGLT-2 inhibitors, such as canagliflozin, dapagliflozin, and empagliflozin, have been explored successfully against type II diabetes [6 - 8]. Pro-Xylane™, a *C*-alkyl

* **Corresponding author Nazar Hussain:** Department of Medicinal Chemistry, Institute of Medical Sciences, Banaras Hindu University, Varanasi-221005, India; E-mail: nazar10@bhu.ac.in

glycoside, has been explored as a popular skin anti-aging agent. The *C*-analogs of KRN7000 and blood group H-antigen are known to exhibit various biological activities, including anticancer activity, as depicted in Fig. (**1**) [9 - 11].

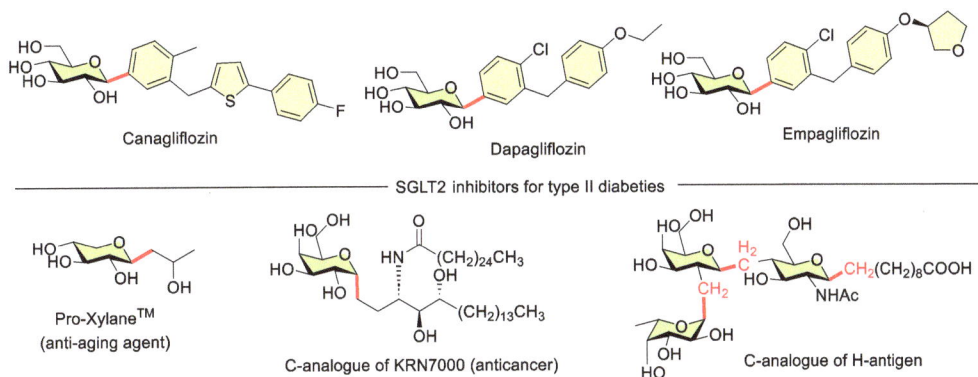

Fig. (1). Selected examples of metabolically stable *C*-glycoside drugs.

Moreover, aryl *C*-glycosides are widely distributed in several bioactive natural products and popular drug candidates. The natural products derived from aryl *C*-glycosides have particularly attracted a great deal of attention due to their potential bioactivity ranging from anti-inflammatory to anticancer, as depicted in Fig. (**2**). On the other hand, 2-Deoxy-β-*C*-glycosides embody an important class of carbohydrate derivatives and are widely distributed in various bioactive motifs *e.g.*, saptomycin B and vieomycinone B$_2$ methyl ester (Fig. **2**).

In this regard, the research group of Keisuke Suzuki has demonstrated the synthetic challenges of aryl *C*-glycoside natural products in their preparation [5, 12]. The authors covered two aspects in this review: (i) synthetic approaches and their application for the total synthesis of aryl *C*-glycoside natural products and (ii) synthetic strategies that favor the glycosylation of arenes utilizing three different types of reactions, *viz.*, cross-coupling reactions, glycosyl anions, and electrophilic sugar derivatives. A number of glycosyl donors such as glycals, 1,2-anhydro sugars, sugar lactols and lactones, glycosyl phosphates/imidates, methyl glycosides, glycosyl acetates, chalcogenoglycosides/sulfoxides/sulfones, glycosyl halides, and other glycal derivatives are archetypally used in the *C*-glycosylation reactions, as illustrated in Fig. (**3**).

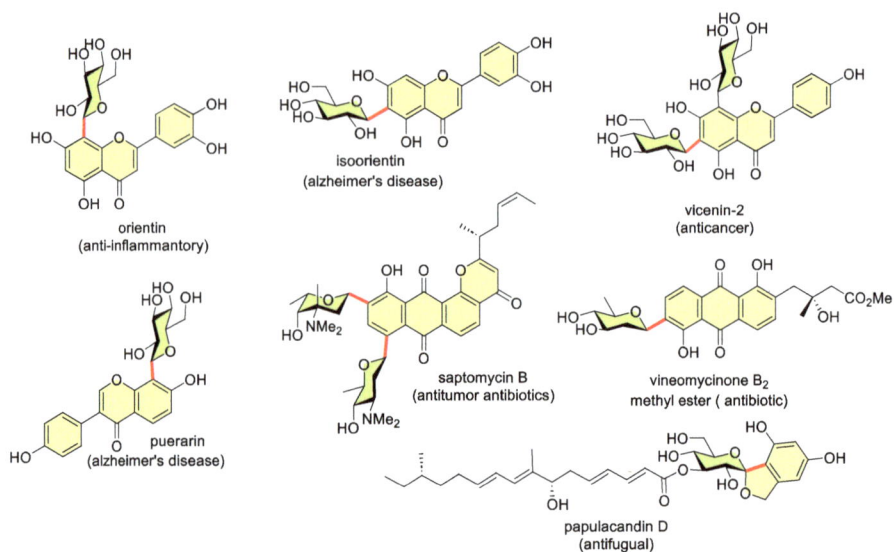

Fig. (2). Some selected examples of biologically active aryl *C*-glycoside natural products.

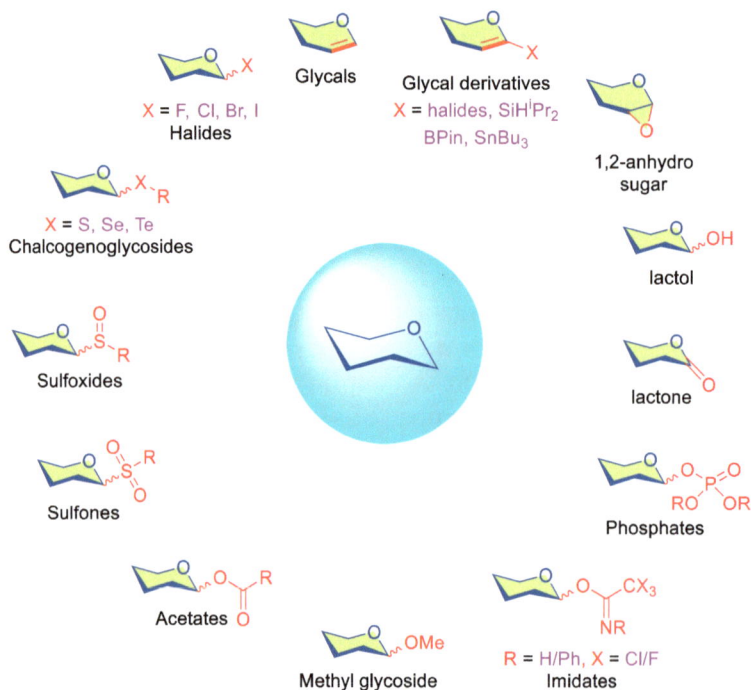

Fig. (3). Various glycosyl donors employed in the synthesis of *C*-glycosides.

In 2019, Kinfe reviewed the literature on different types of transformations of *endo* glycals into various biologically relevant motifs, including bergenin, papulacandin D, sphinganine, diospongins, decytospolides, thiophene *C*-glycosides, isatisine, phomonol, thailanstatin A methyl ester, *etc.*, as shown in Fig. (**4**) [13].

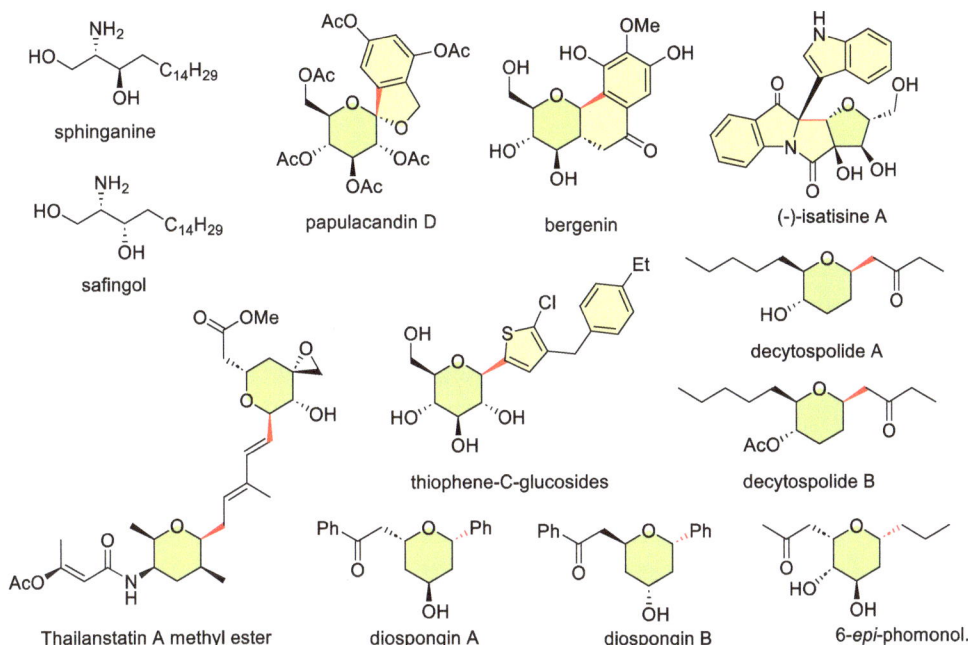

Fig. (4). Selected examples of biologically relevant motifs derived from *endo* glycals.

SYNTHESIS OF *C*-GLYCOSIDES USING GLYCALS AND THEIR DERIVATIVES

Glycals are widely used as an electrophilic cationic sugar species for creating a new anomeric C-C bond leading to the construction of *C*-glycosides. The most adopted methods for the synthesis of *C*-glycosides include Ferrier-type *C*-glycosylation, transition-metal-catalyzed *C*-glycosylation, and other types of *C*-glycosylation using glycals and their derivatives. Nowadays, a number of efforts have been made to devise other valuable synthetic strategies to access *C*-glycosides involving photo-redox catalysis.

Ferrier Type *C*-glycosylation

The glycosylation reaction *i.e.*, the formation of a glycosidic bond, is a fundamental reaction in carbohydrate chemistry. Generally, the electrophilic *C*-

glycosylation of glycals using *C*-nucleophiles proceeds through Ferrier rearrangement in the presence of stoichiometric amounts of Lewis acids or promotors, favoring the formation of α-anomers, as shown in (Scheme **1**).

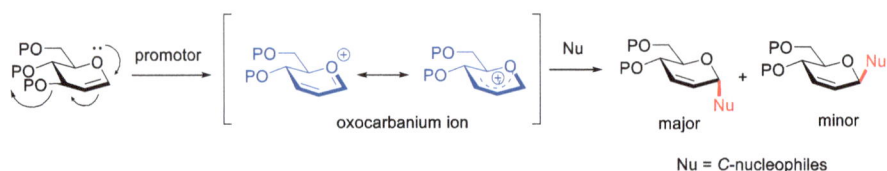

Scheme 1. General mechanistic pathway for the synthesis of *C*-glycosides *via* Ferrier rearrangement.

For instance, Huang and co-workers have described the synthesis of C1-xylopyranosides **3** employing Ferrier rearrangement/intramolecular 1,3-acyloxy migration strategy by making use of D-xylal **1** and propargylic carboxylates **2** in a β-selective manner, as illustrated in (Scheme **2**) [14]. The authors disclosed that the combination of Ph$_3$PAuCl and AgSbF$_6$ was an effective catalytic system for the reaction. The method opened up a new prospect for the synthesis of value-added β-selective *C*-glycosides.

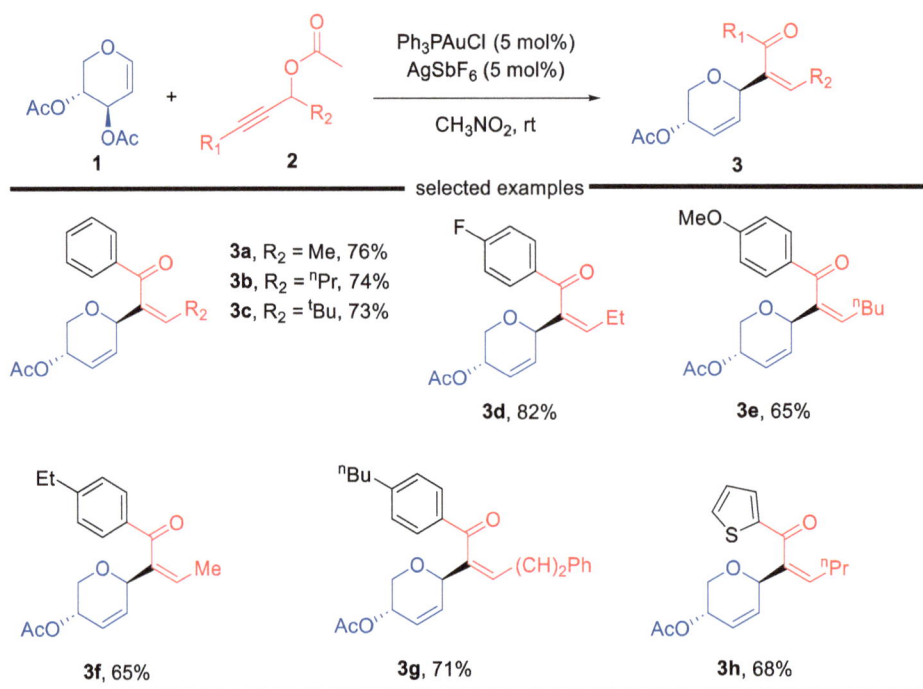

Scheme 2. Ph$_3$PAuCl/AgSbF$_6$ catalytic system for the synthesis of β-selective *C*-glycosides.

A plausible mechanism for the synthesis of compound **3** is outlined in (Scheme **3**). The reaction is assumed to proceed *via* the formation of allylic oxocarbenium ion **A** from D-xylal **1** through a Ferrier rearrangement. Further, the propargylic carboxylate **2** is transformed into the nucleophilic allenic intermediate **B**, which attacks the electrophilic anomeric carbon of the intermediate **A,** followed by 1,3-acyloxy migration leading to the formation of oxocarbenium ion **C**. Finally, the intermediate **C** delivers the target product **3** through hydrolysis.

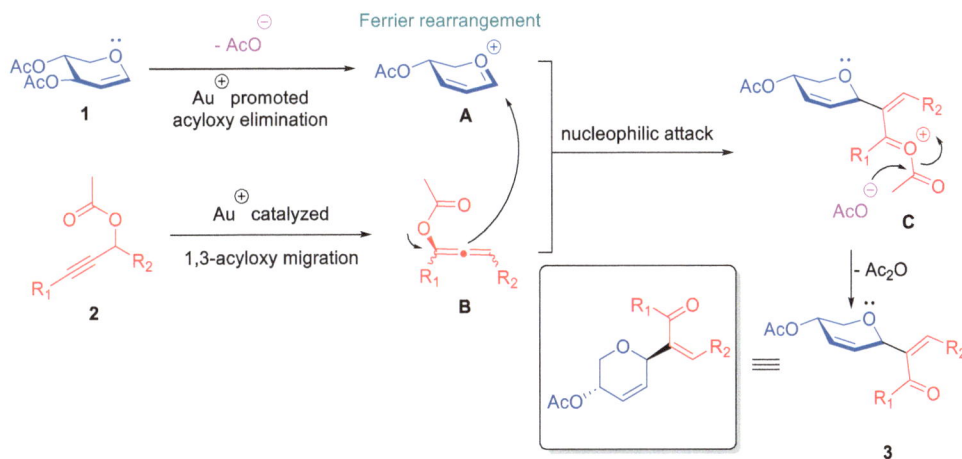

Scheme 3. Plausible mechanism for the synthesis of β-selective *C*-glycosides.

Another approach was adopted for the diastereoselective synthesis of aryl vinyl *C*-glycosides (**8-10**) using various glycals **4** and allyl silanes (**5-7**) through Ferrier rearrangement by the research group of Kancharla, as illustrated in (Scheme **4**) [15]. The authors used a bulky and strained protonated 2,4,6-tri-*tert*-butylpyridine (TTBPy) triflate as an effective organocatalyst in this method. Several protecting groups on glycals, *viz.*, –OAc, –OBn, –OBz, and –OTBDPS, were well tolerated during the reaction. In the case of silyl enol ethers **6**, β-selective products (**9a-c**) were obtained, while α-selective products (**8a-g & 10a-c**) were isolated when cinnamylsilanes **5** and allyltrimethylsilanes **7** were used, respectively.

Transition-metal (TM)-catalyzed *C*-glycosylation

In recent years, TM-catalyzed *C*-glycosylation has emerged as a research hotspot in the carbohydrate chemistry for synthesizing *C*-glycosides. The usage of TM complexes in catalytic amounts for glycosylations curtails the production of chemical waste, which is typically a problem in Lewis acids or other promoter-based glycosylation reactions where stoichiometric amounts are required. Additionally, stereoselective glycosylations can be achieved by tuning the stereoelectronic nature of the ligands already attached to the transition metals or through the coordination of transition metals to the heteroatoms like *O & N* of glycosyl donors and acceptors. Furthermore, the advancement toward several TM-catalyzed chemoselective activations of anomeric leaving groups has augmented the "toolbox" for orthogonal glycosylation tactics.

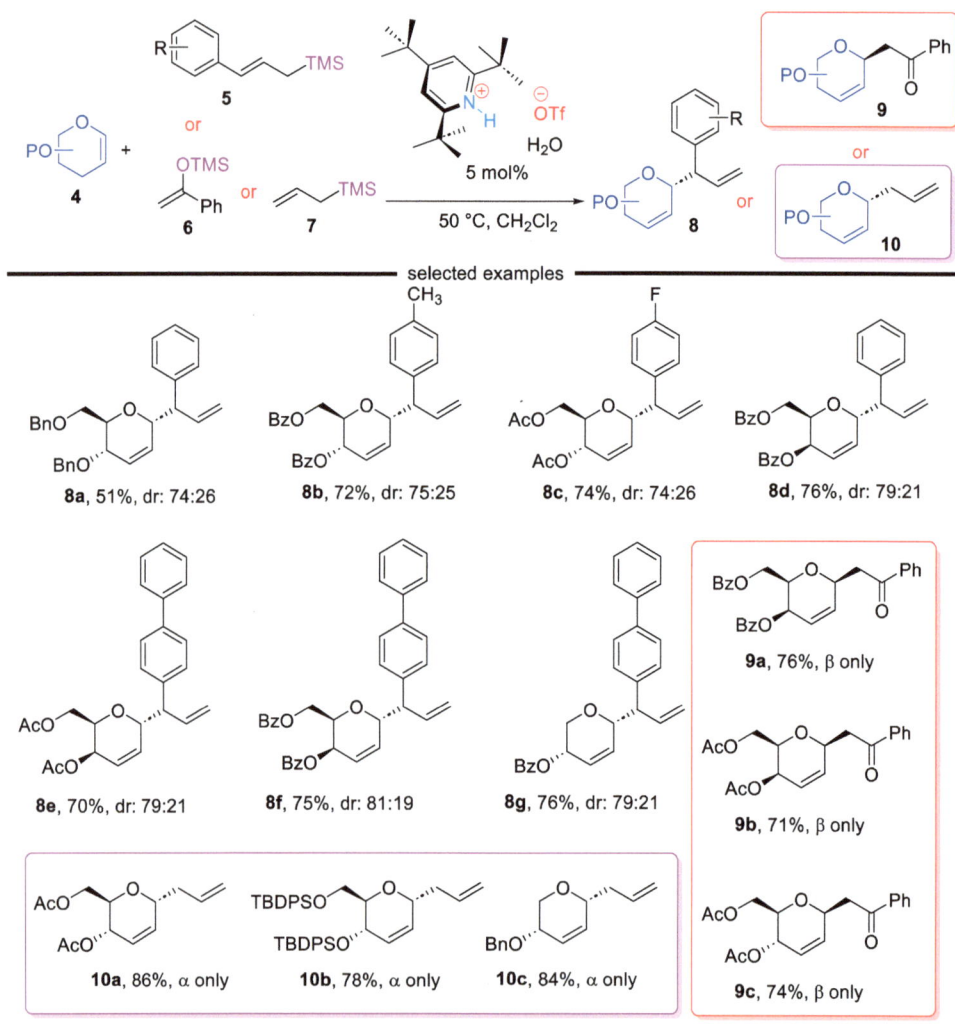

Scheme 4. TTBPy triflate promoted synthesis of phenylallyl Ferrier glycosides from glycals.

Palladium-catalyzed C-glycosylation

Recently, Wei *et al.* have reported a convenient method for the synthesis of aryl *C*-glycosides **12** *via* Pd(II)-catalyzed Heck reaction of non-activated glycals **4** and aryl thianthrenium salts **11**, as depicted in (Scheme **5A**) [16]. A number of protected glycal derivatives such as D-glucal, D-galactal, and L-rhamnal were involved in the reaction to afford respective products in good yields. α-selective products (**12a-g**) were formed in the case of D-glucal and D-galactal; however, L-rhamnal provided β-selectivity (**12h**). The method allows the incorporation of diverse glycals into structurally varied aglycone units without directing groups or pre-functionalization, offering a new avenue for the production of complex *C*-glycosides. The versatility of the method was further demonstrated by the late-stage *C*-glycosylation with numerous natural products (*e.g.*, etofenprox, **12d**).

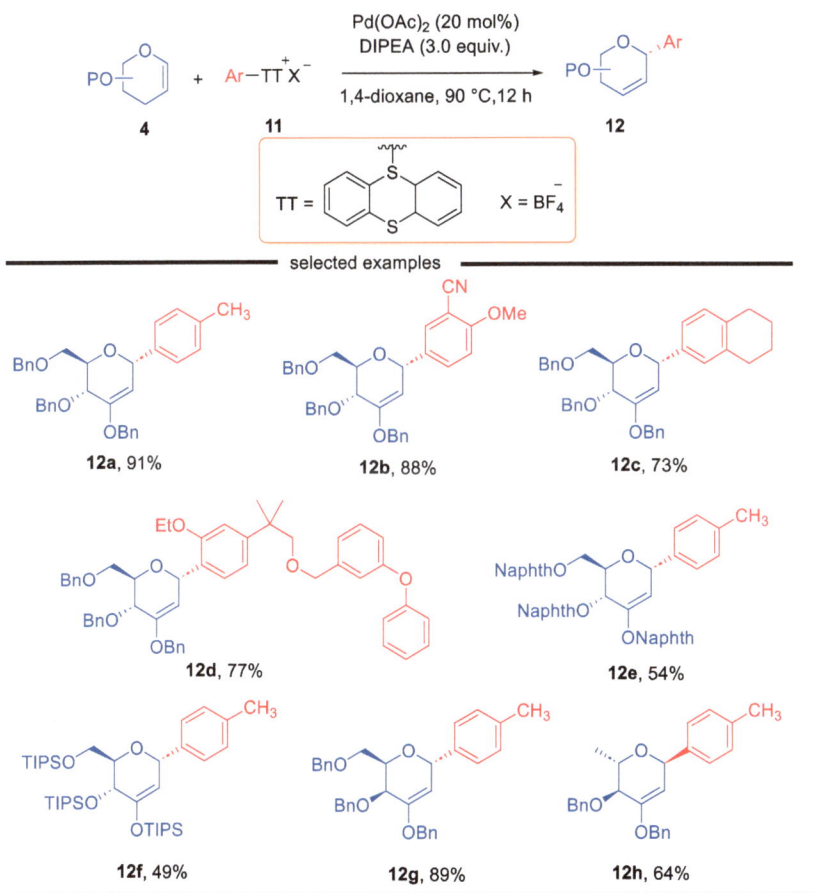

Scheme 5A. Pd-catalyzed synthesis of aryl *C*-glycosides from glycals and aryl thianthrenium salts.

A proposed mechanistic pathway for the preparation of aryl *C*-glycosides **12** is depicted in (Scheme **5B**). The reaction is assumed to proceed *via* the reduction of Pd(II) into Pd(0) through ligand exchange and elimination of acetic acid. The active Pd(0) species **I** is formed and then enters into the catalytic cycle to produce the intermediate **II** *via* oxidative addition of **11**. Further, a glycal-Pd(II) complex **III** is produced after the carbopalladation with **4,** and the complex **III** then undergoes β-hydride elimination to yield the target product **12** along with the generation of intermediate **IV**. Finally, the intermediate **IV** undergoes reductive elimination to regenerate the initial Pd(0) catalyst **I,** which will be ready for the catalytic cycle.

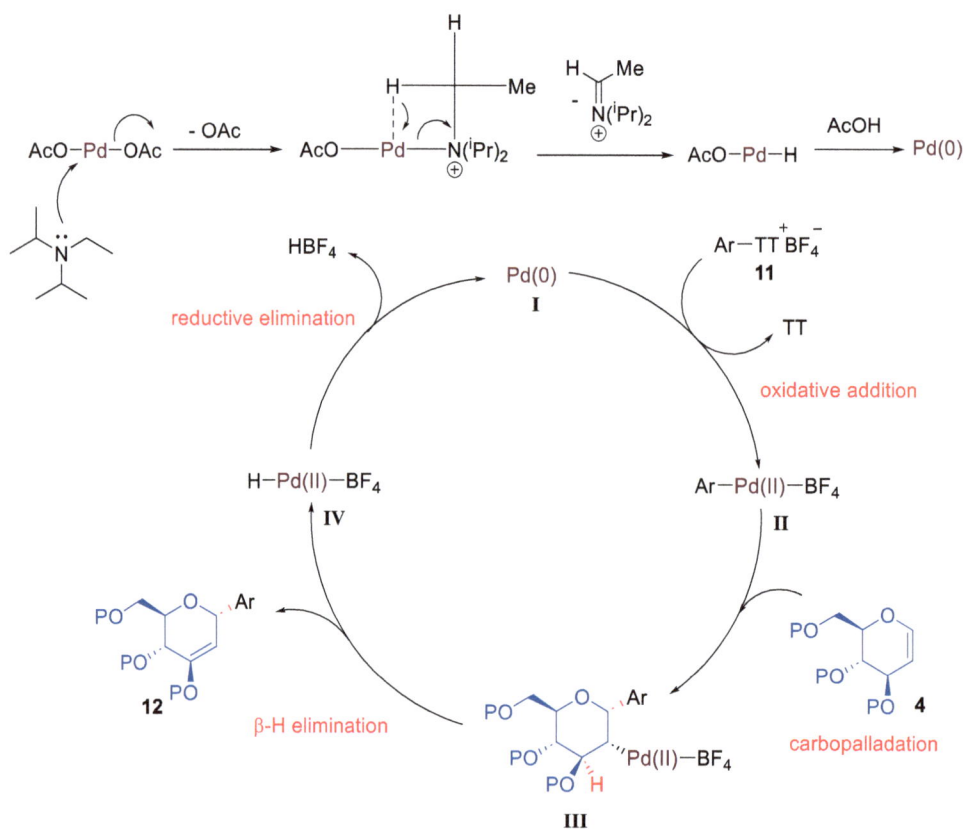

Scheme 5B. Mechanistic pathway for the Pd-catalyzed synthesis of aryl *C*-glycosides.

Pal *et al.* disclosed a competent method for the stereoselective synthesis of 2,3-dideoxy *C*-aryl glycosides **14** *via* Pd(II)-catalyzed *C*-glycosylation of glycals **4** with diaryliodonium salts **13**, as depicted in (Scheme **6**) [17]. A number of substituents such as Me, OMe, Br, COMe, and CF$_3$ on the aryl ring of diaryliodonium salts were involved during the course of the reaction to produce the desired products (**14a-f**). A naphthyl-substituted diaryliodonium triflate reacted with 3,4,6-tri-*O*-acetyl-D-glucal, affording the corresponding *C*-glycoside **14g** in 53% yield.

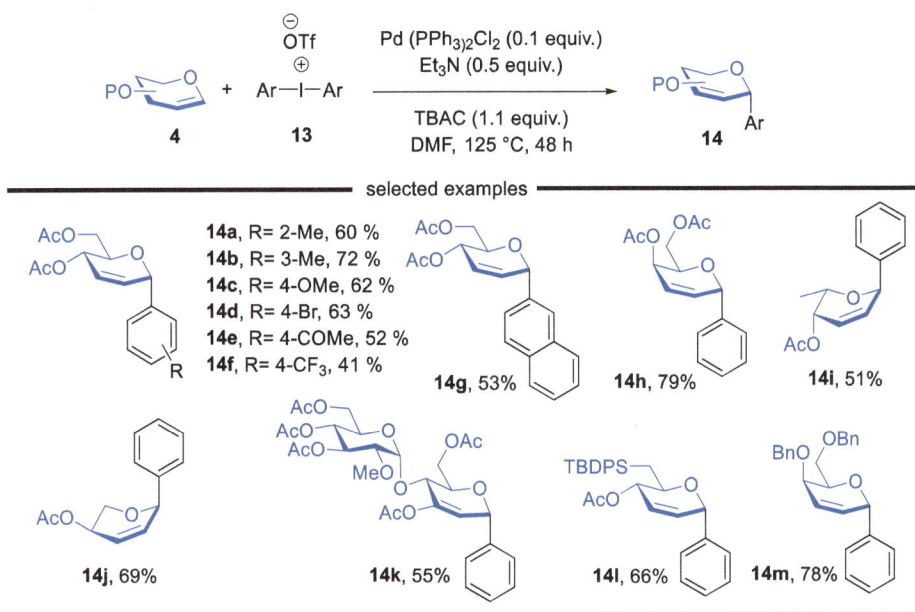

Scheme 6. Pd-catalyzed synthesis of aryl *C*-glycosides using glycals and diaryliodonium salts

Numerous protected glycals, *viz.*, D-galactal, D-glucal, L-arabinal, L-fucal, and D-maltal, were tolerated well to give glycosylated products (**14h-m**). This protocol has been praised as a noteworthy contribution to the field of carbohydrate chemistry owing to its good functional group tolerability and broad substrate scope.

The research group of Hussain has developed a Pd-catalyzed protocol for the stereoselective synthesis of chromone *C*-glycosides **16** using glycals **4** and 3-halo chromones **15**, as illustrated in (Scheme 7) [18]. A 3,4-dimethylated glucal containing different substituents at the primary hydroxyl group, such as H, Me, TBS, and Ac, successfully endured the reaction, affording the products **16a-d** in good yields. Moreover, numerous pharmaceuticals, including modified derivatives of oleic acid, ciprofibrate, and ibuprofen, linked with D-glucal, were effectively transformed into the anticipated chromone *C*-glycosides **16e-g**. The authors extended this study to glycals other than glucals, including those derived from D-galactose and L-rhamnose, which produced glycosylated products **16h** and **16i** in good yields. The substrate scope was further broadened using methylated D-maltal, which gave the corresponding product **16j** in 57% yield. A number of chromone derivatives containing various substituents such as Cl, Me, OMe, and naphthyl were involved in the reaction, affording the corresponding *C*-glycosides **16k-o** in good yields ranging from 61-74%.

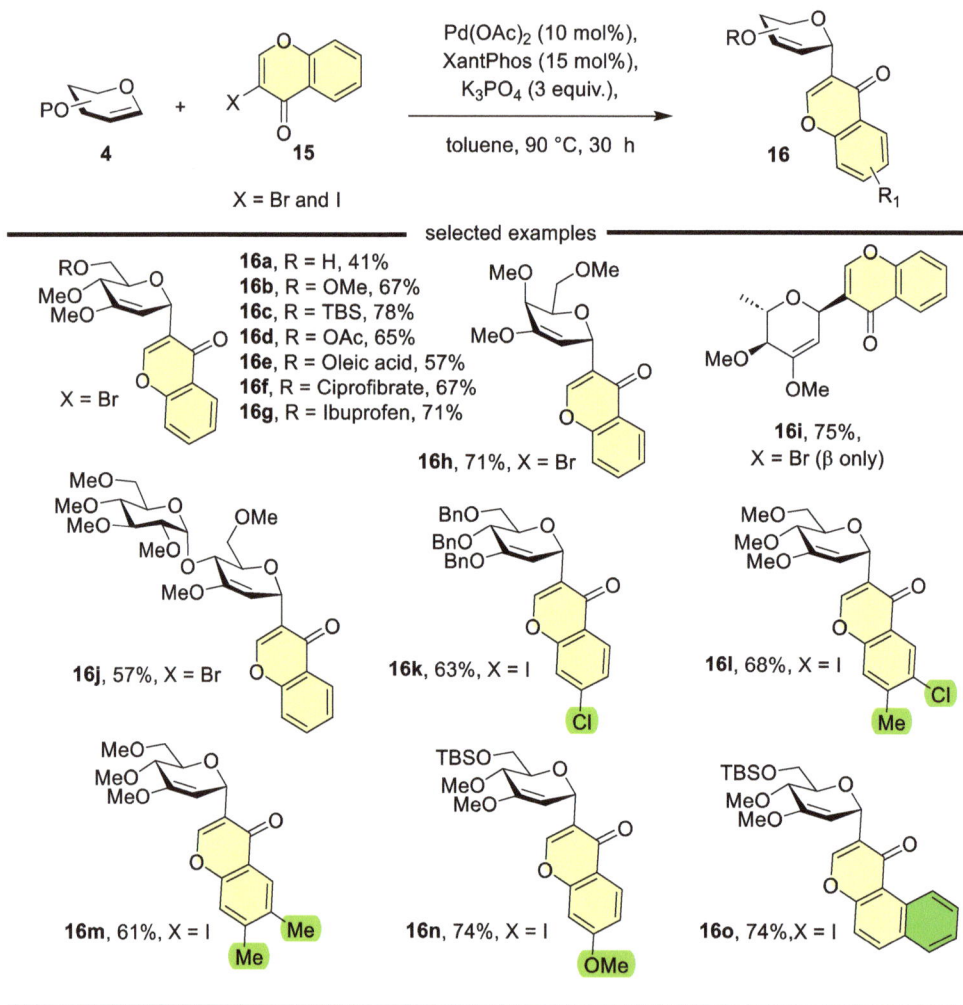

Scheme 7. Pd-catalyzed synthesis of chromone *C*-glycosides.

The research group of Kandasamy has developed a competent method for the stereoselective synthesis of C-1 aryl enones **19** from glycal enones **17** and arylboronic acids **18** using Pd-catalysis, as depicted in (Scheme **8**) [19]. The authors disclosed that a number of enones derived from benzyl-protected D-glucal, D-galactal, L-rhamnal, D-rhamnal, and L-arabinal participated in the reaction with arylboronic acids having various substituents, *viz.*, Me, OMe, halides, CN, CF_3, and NO_2 on the aryl ring, giving rise to several C-1 aryl enones (**19c-h**, **19i-l**, **19r**, **19s**, and **19t**) in good yields. Furthermore, numerous protecting groups in D-glucal enones (such as Ac, Bz, Piv, and MOM) were found to be well tolerated, affording the desired products **19n-q** in good yields. However, D-glucal and galactal-enone reacted with naphthelene-1-boronic acid under the standard conditions to afford the products **19a** and **19m**, respectively.

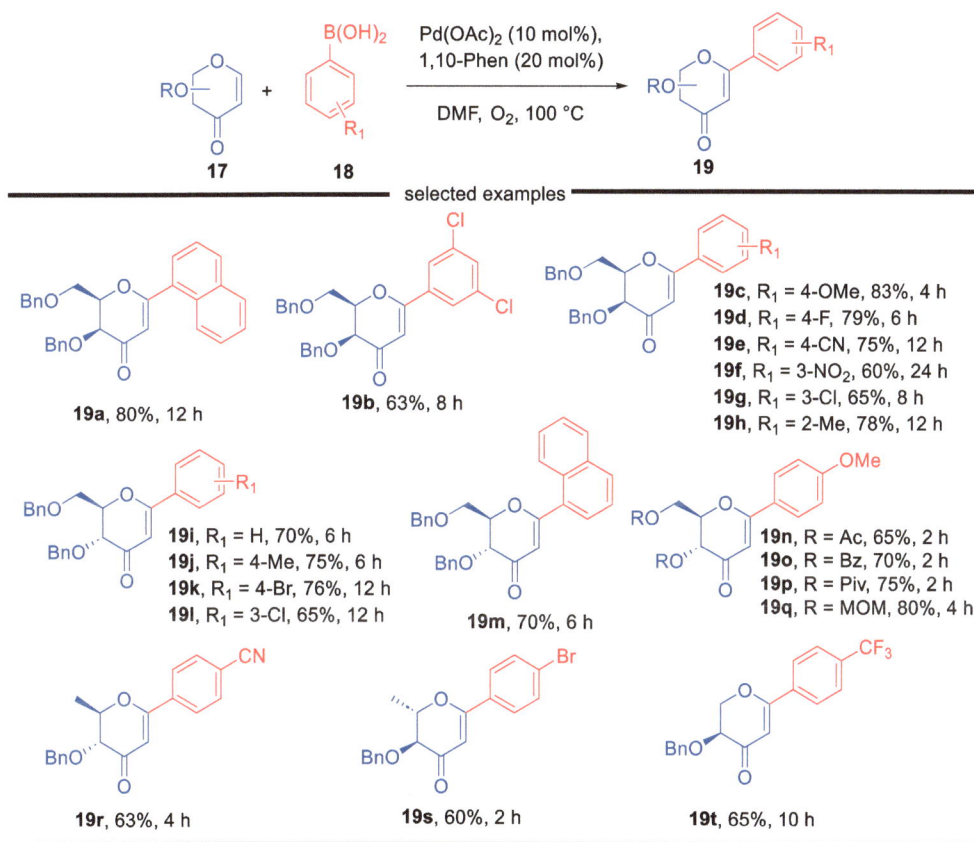

Scheme 8. Pd-catalyzed stereoselective synthesis of C-1 aryl enones.

The research group of Mandal has disclosed a directed $C(sp^2)$-H functionalization at the anomeric position on C2-amidoglycals **20** with various *p*-quinone methides (*p*-QMs) **21** to afford *gem*-diarylmethyl *C*-glycosides **22** using Pd-catalysis, as illustrated in (Scheme **9**) [20]. The aryl ring of *p*-QMs having various substituents like Me, OMe, NMe$_2$, F, and Br reacted efficiently with peracetylated-2-amidoglucal, affording the corresponding glycosylation products **22a-f** in 73-87% yields. Apart from this, tri-and di-substituted *p*-QMs also reacted proficiently to give the products **22g** and **22h** in 89% and 79% yields, respectively. Further, the reaction of indole and thiophene-based *p*-QMs with glycals delivered the products **22i** and **22m**, respectively. Additionally, the glycal structures (*e.g.*, D-galactal, L-rhamnal, and D-lactal) showed similar reactivity in this C-H functionalization with different *p*-QMs, leading to the products (**22j-l** and **22n**) in good yields.

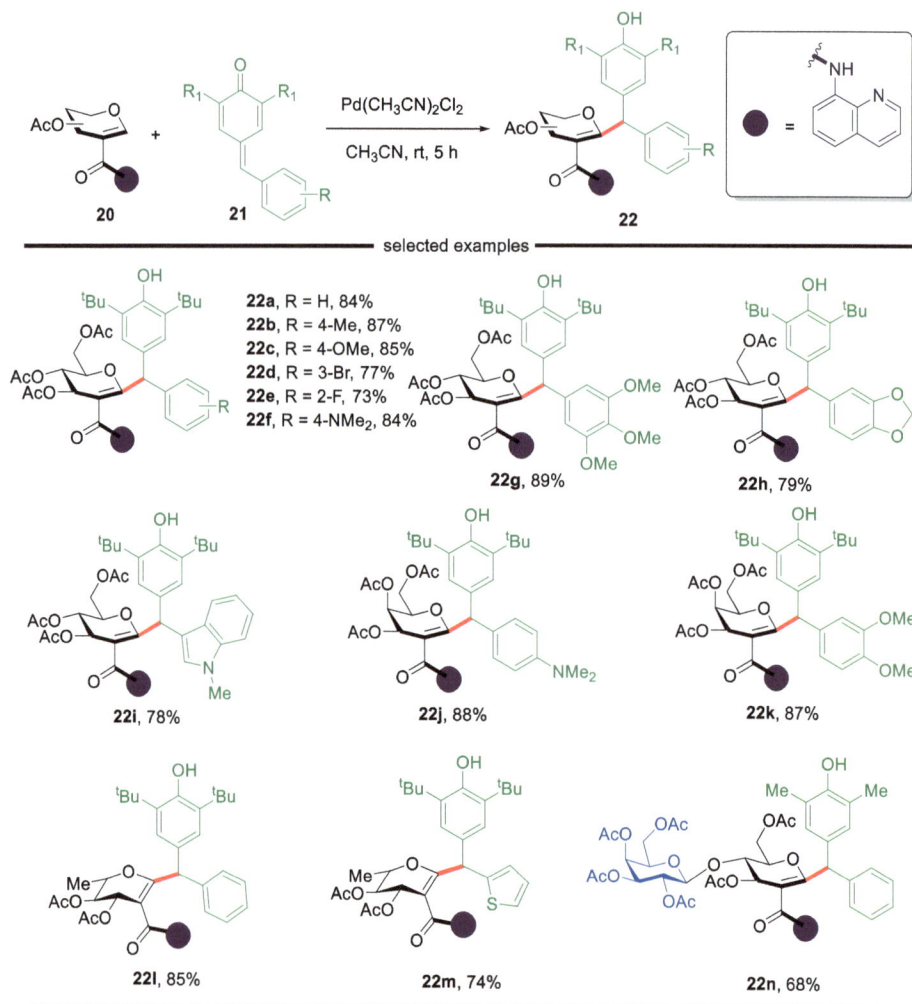

Scheme 9. Pd-catalyzed synthesis of *gem*-diarylmethyl *C*-glycosides.

Xiao *et al.* disclosed an effective approach for the stereoselective preparation of indolyl-*C*-glycosides **24** *via* sequential aminopalladation and Heck glycosylation of 2-alkynylanilines **23** with glycals **4**, as depicted in (Scheme **10**) [21]. The authors demonstrated that the different protections on D-glucal moiety were tolerated well and transformed into corresponding indolyl-*C*-glycosides **24a-b** in 76-61% yields. 3,4,6-tri-*O*-acetylgalactal reacted with 2-alkynylanilines **23** bearing Me and Cl groups on the phenyl ring, giving rise to the desired products **24c** and **24d** in 81% and 67% yield, respectively. Moreover, peracetylated L-rhamnal afforded the products **24e-f** with excellent yield and stereoselectivity.

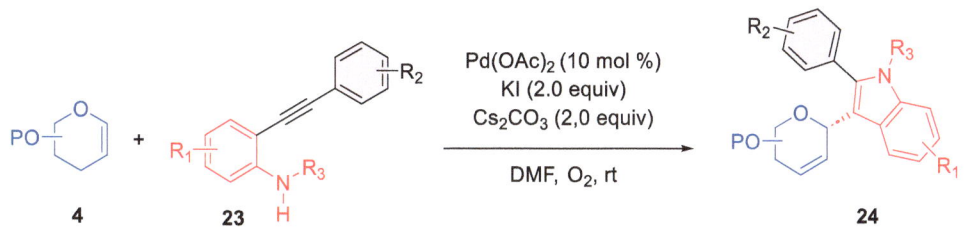

selected examples

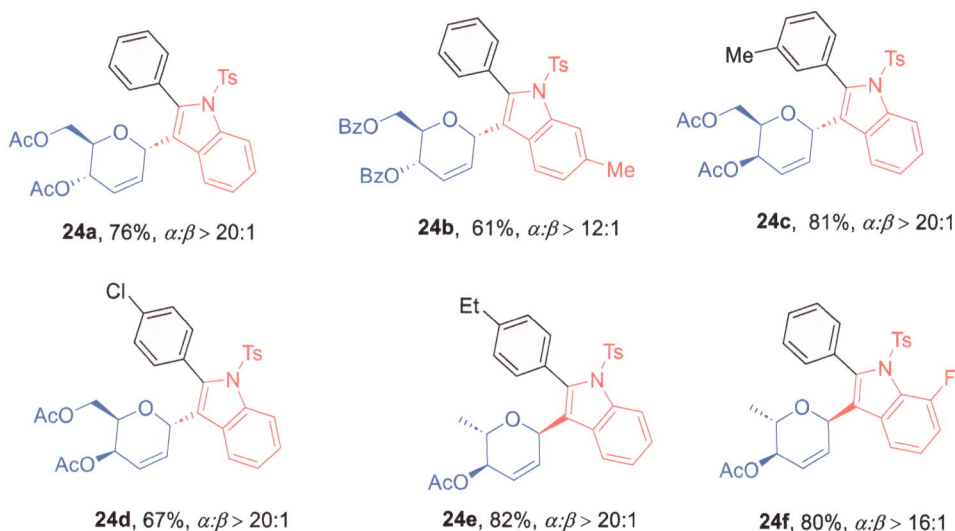

24a, 76%, $\alpha{:}\beta > 20{:}1$ **24b**, 61%, $\alpha{:}\beta > 12{:}1$ **24c**, 81%, $\alpha{:}\beta > 20{:}1$

24d, 67%, $\alpha{:}\beta > 20{:}1$ **24e**, 82%, $\alpha{:}\beta > 20{:}1$ **24f**, 80%, $\alpha{:}\beta > 16{:}1$

Scheme 10. Pd-catalyzed synthesis of indolyl-*C*-glycosides.

A plausible mechanistic cycle for synthesizing indolyl-*C*-glycosides is depicted in (Scheme **11**). Firstly, Pd(OAc)$_2$ reacts with KI to form the active PdI$_2$ catalyst **I**, which could coordinate with **23** to afford the intermediate **II**. The consequent intramolecular *trans*-nucleopalladation of the alkyne creates an indolylpalladium intermediate **III**, which undergoes regioselective migratory insertion across the double bond of glycal **4** to form intermediate **IV**. Finally, the intermediate **IV** undergoes trans elimination to afford the target product **24** along with the active PdI_2 catalyst **I**, which is ready for the next catalytic cycle.

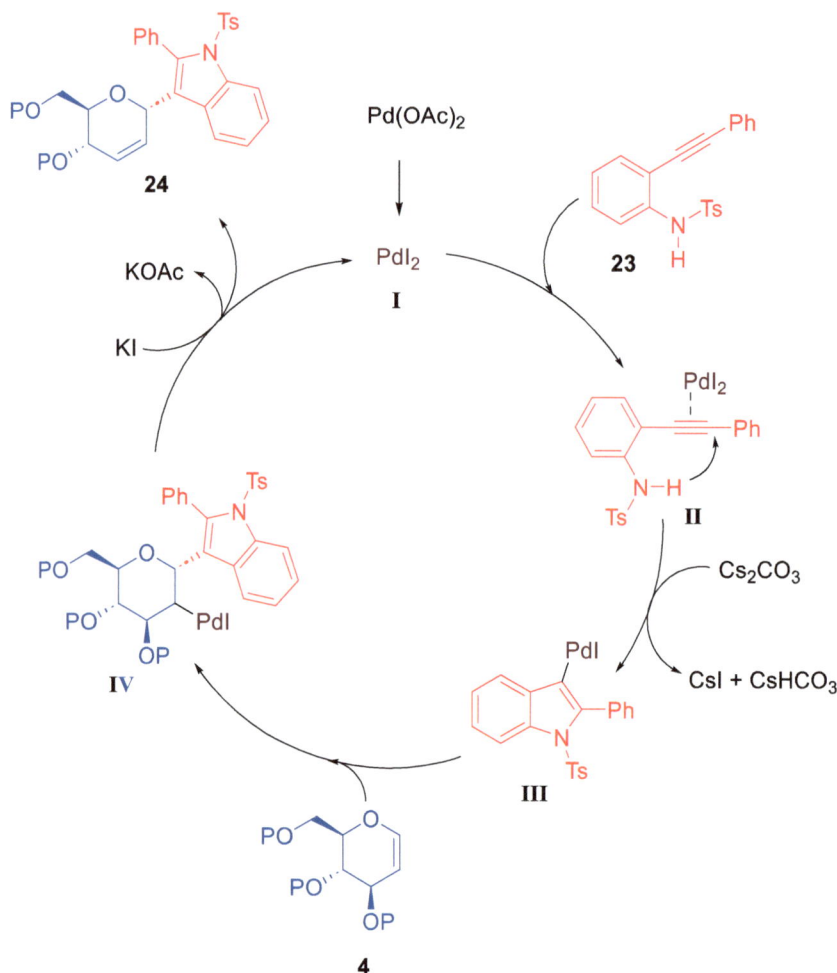

Scheme 11. Plausible reaction pathway for the Pd-catalyzed synthesis of indolyl-*C*-glycosides.

Liu *et al.* have disclosed a practical method for the synthesis of 3-indolyl $C\text{-}\Delta^{1,2}$-glycosides **26** using Pd-catalyzed annulation/*C*-glycosylation of *o*-alkynylanilines **23** with 1-iodoglycals **25**, as illustrated in (Scheme **12**) [22]. The method features a wide substrate scope with variation in reactant **23** having different substituents on indole ring and phenyl moiety, leading to 3-indolyl-*C*-$\Delta^{1,2}$-glycosides **26a-e** and **26f-g**, respectively. Furthermore, the reactions of L-rhamnal with **23** delivered desired *C*-glycoside products **26h** (93%) and **26i** (79%). The authors also examined the protocol with *N*-protected *o*-alkynylanilines for the preparation *C*-glycosides **26j** in 74% yield.

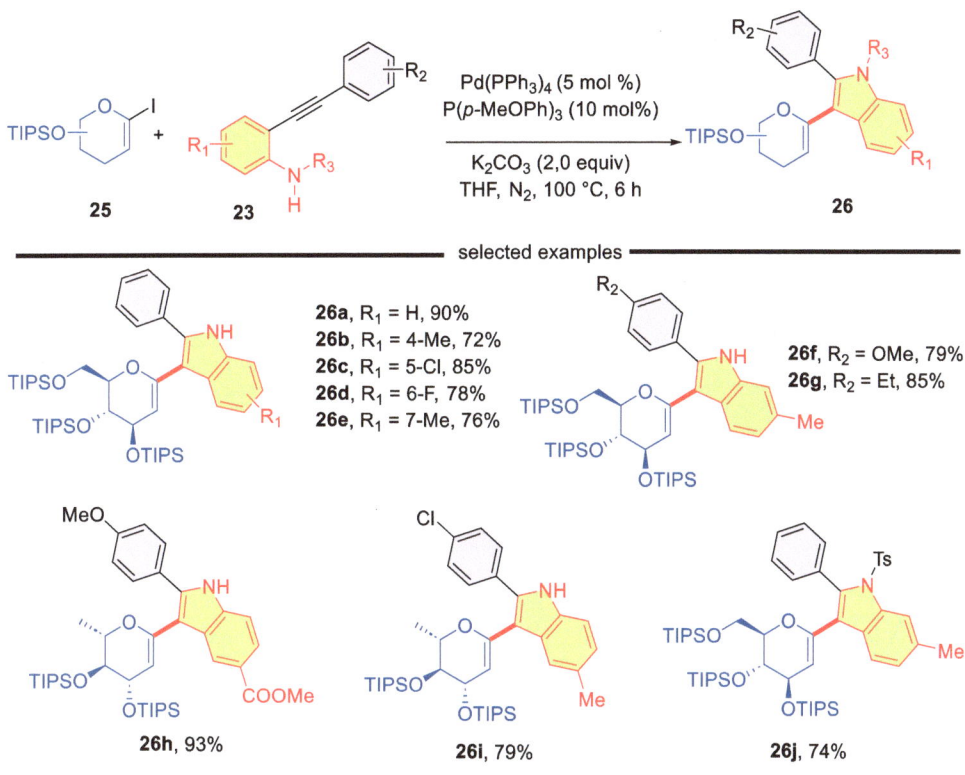

Scheme 12. Pd-catalyzed synthesis of 3-indolyl *C*-$\Delta^{1,2}$-glycosides.

The research group of Ackermann has devised a Pd-catalyzed approach for the late-stage C(sp^3)-H glycosylation of 2-deoxy β-glycosides **27** with 1-iodoglycals **25** to afford structurally diverse *C*-oligosaccharides **29**, as depicted in (Scheme **13**) [23]. The strategy is operationally simple, efficient, and diastereoselective and provides a remarkable prospect for accessing various *C*-oligosaccharides. The author summarized that 2-deoxy-glucosides and galactosides bearing different protecting groups such as Ac, Bz, Me, and Bn delivered the anticipated glucosyl and galactosyl-*C*-[2→1]-glucal **29a-d** and **29e-g** in good to excellent yields, respectively. ^{1}C$_4$-conformation of *C*-glycosides were used for the *trans*-selective C(sp^3)-H glycosylation to afford rhamnosyl-*C*-[2→1]-glucal **29h** and fucosyl-*C*-[2→1]-galactal **29i** with 64% and 76% yields, respectively. In addition, various 2-deoxy-*C*-glycosides derived from galactose, lactose, maltose, and cellobiose reacted with TIPS-protected 1-iodoglucal to give respective *C*-oligosaccharides

29j and **29n-p** in moderate to good yields. Further, the glycosyl donors derived from rhamnose and galactose were also found to be amenable to 2-deoxy-glucosides and galacotosides, affording disaccharides **29k-l** and **29m** in moderate to excellent yields.

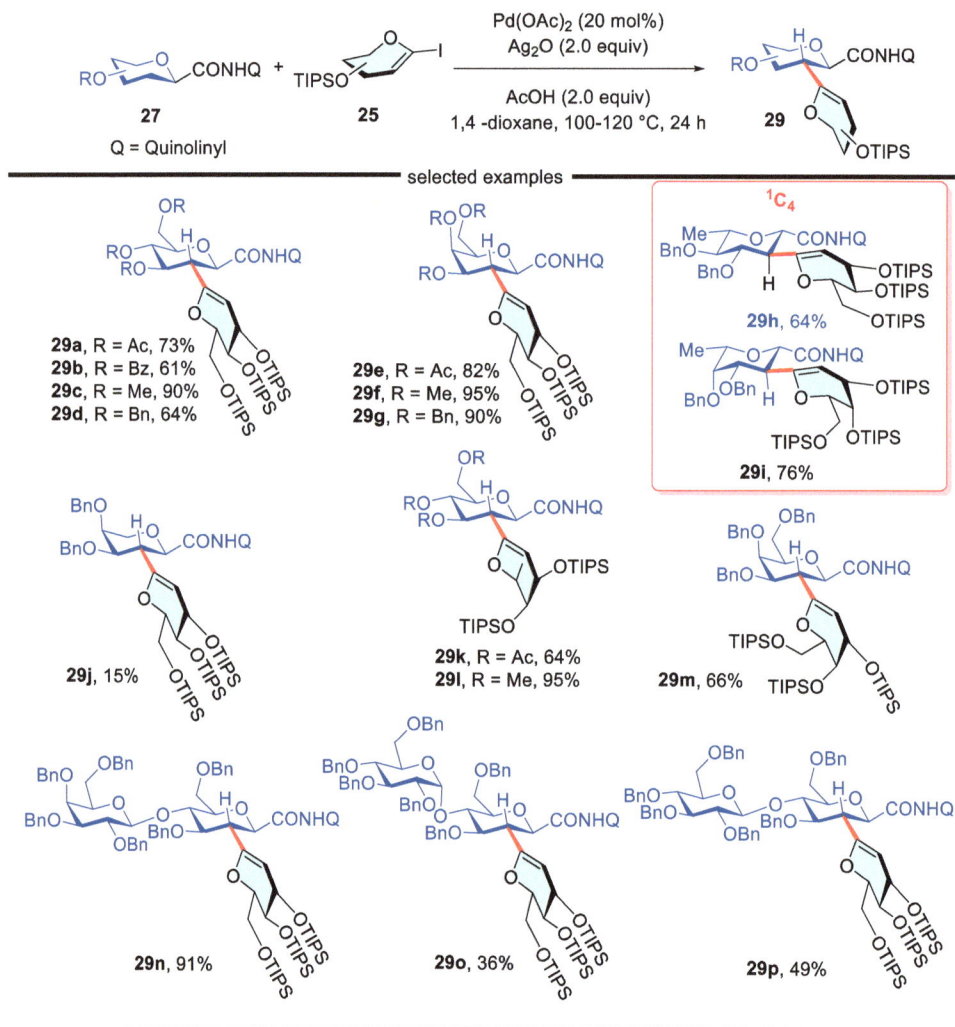

Scheme 13. Pd-catalyzed synthesis of *C*-oligosaccharides.

Xiang and co-workers have described a Pd-catalyzed decarboxylative reaction for the β-selective *C*-glycosylation of bicyclic galactals **30** with 2-oxindoles **31** to access various *C*-glycosides **32** under mild reaction conditions, as illustrated in

(Scheme **14**) [24]. The authors disclosed that the nucleophilicity of 2-oxindole was increased *in situ* due to the deprotonation of the enol tautomer by the decarboxylation intermediate of galactal. The C3 position of 2-oxindole **31** containing methyl ester group reacted with glycal **30,** resulting in *C*-glycoside **32a** with 93% yield. However, the products **32b** and **32c** were obtained in 96% and 81% yields, respectively, when 2-oxindole **31** bearing Me and OMe groups were at the C5 position. Further, the *N*-phenyl-protected 2-oxindole gave the desired β-selective product **32d** in 93% yield. The authors demonstrated the potential of this method by examining various protecting groups such as TIPS, TBDPS, Bn, Ac, and Piv on C6-oxygen of bicyclic galactals, affording β-*C*-glycosides **32e-i** with moderate to good yields.

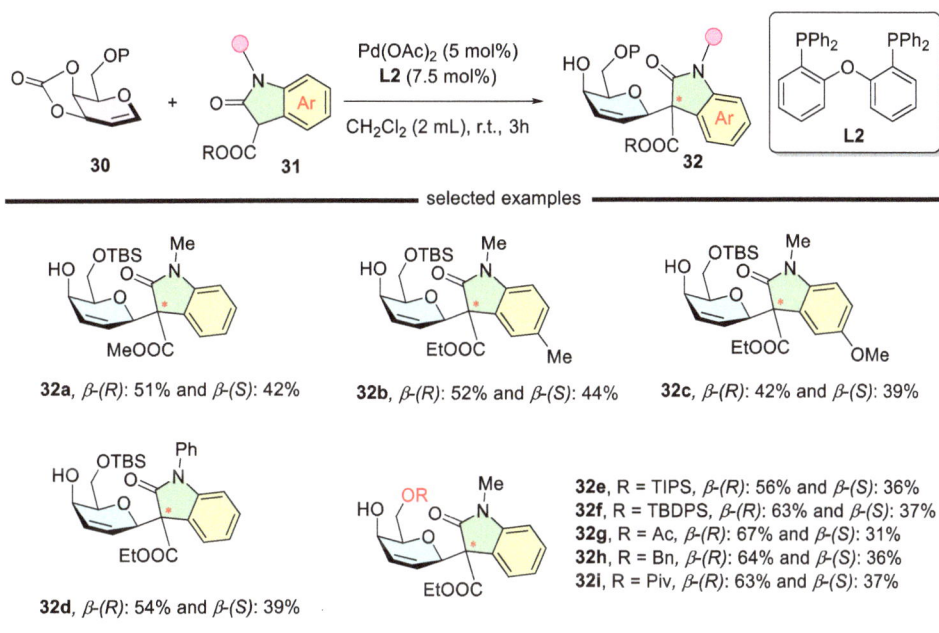

Scheme 14. Pd-catalyzed synthesis of aryl β-selective *C*-glycosides.

The research group of Parkan has disclosed a Pd-catalyzed Hiyama reaction of unprotected 1-diisopropylsilyl-D-glucal **33** with iodo- or bromo-(hetero) arenes **34** for synthesizing various (hetero)aryl *C*-glycosides **35** under mild conditions, as demonstrated in (Scheme **15**) [25]. The unprotected 1-diisopropylsilyl-D-glucal **33**, on reaction with 1-halonaphthalene, delivered the desired aryl *C*-glycoside **35a** in 81% yield. A trimethoxy-substituted haloarene reacted with 1-diisopropylsilyl-D-glucal **33** to afford the desired product **35b** in 89% yield. A number of electron-withdrawing substituents, *viz.*, OBn, NO$_2$, COCH$_3$, and Br, on the aryl ring of aryl halides participated in the reaction, affording the desired products **35c-f**. Thiophene and pyridine-based heteroaryl halides underwent the reaction smoothly to afford the desired products **35g** and **35i** in 70% and 25% yields, respectively.

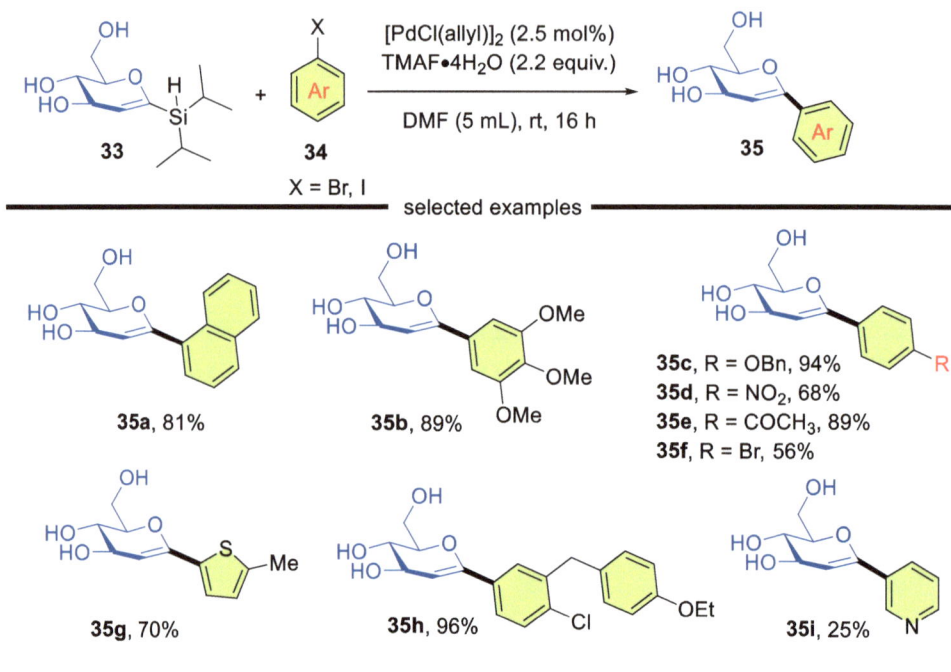

Scheme 15. Pd-catalyzed Hiyama reaction for the synthesis of (hetero)aryl *C*-glycosides.

Liu and coworkers have developed a practical method for the TFA-promoted *N*-quinolylcarboxamide (AQ)-directed glycosylation of inert β-*C*-(sp³)-H bonds of *N*-phthaloyl α-amino acids **36** with glycals **25** to access *C*-alkyl glycoamino acids **37** using Pd-catalysis, as depicted in (Scheme **16**) [26]. A naphthalene-substituted *N*-phthaloyl α-amino acid delivered the desired product **37c** in 66% yield. Besides glucal, the reactions of glycals derived from galactose and rhamnose, with *N*-phthaloyl alanine, also worked well, affording the corresponding products **37e** and **37f** in 71% and 81% yields, respectively. A number of electron-donating, as well as electron-withdrawing substituents such as Me, OMe, Cl, Br, and NO₂, were tolerated well in the reaction to afford the products **37g-l** in good yields. The utility of the method was verified by the effective synthesis of glycopeptides *via* C(sp³)-H activation.

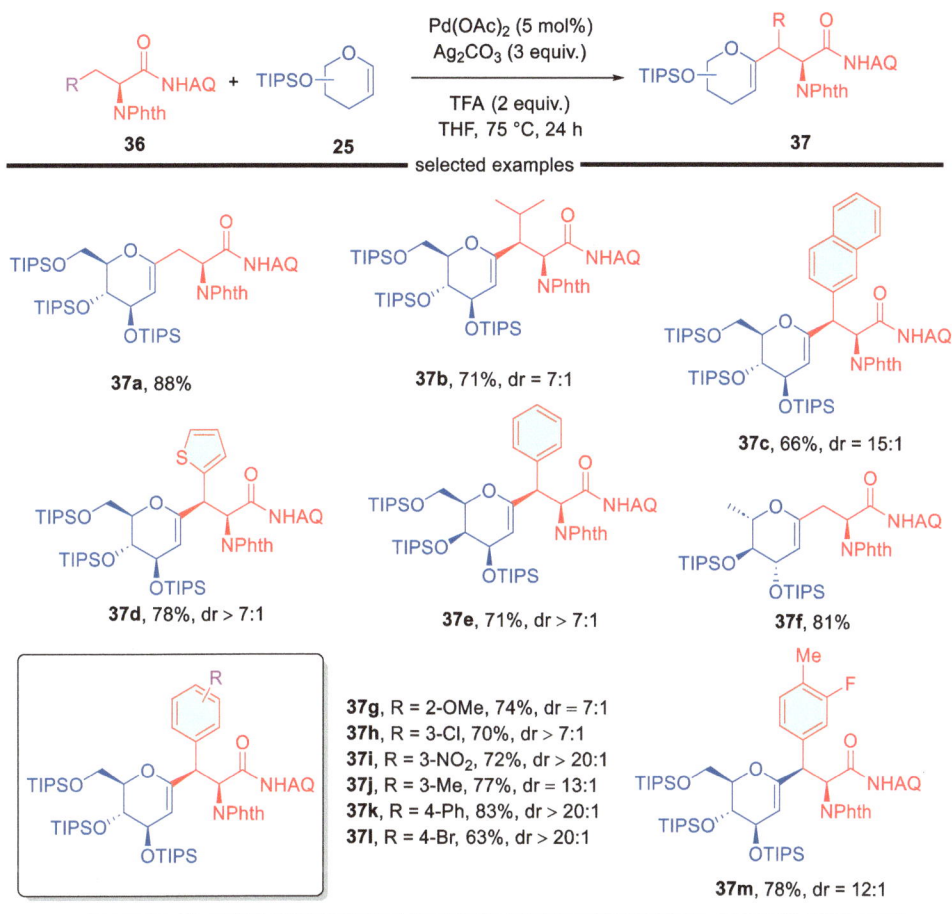

Scheme 16. Pd-catalyzed synthesis of *C*-alkyl glycoamino acids.

Cobalt-catalyzed C-glycosylation

Wang *et al.* have disclosed a ligand-controlled Co(II)-catalyzed $C(sp^3)$-$C(sp^3)$ coupling reaction of glycals **4** with alkyl halides **38** to access β-selective 2-deoxy-*C*-glycosides **39** under mild conditions, as illustrated in (Scheme **17**) [27]. A number of alkyl halides, on reaction with benzyl-protected glucal, afforded the corresponding products **39a-i** in good yields. Other protection on glycals (*e.g.*, glucal and galactal) such as Naphth, PMB, Ac, Me, and Et also tolerated well in the reaction to afford the corresponding products **39j-r** in good yields. The authors disclosed that the reaction follows a radical pathway involving alkyl halide activation and ligand-controlled β-selectivity.

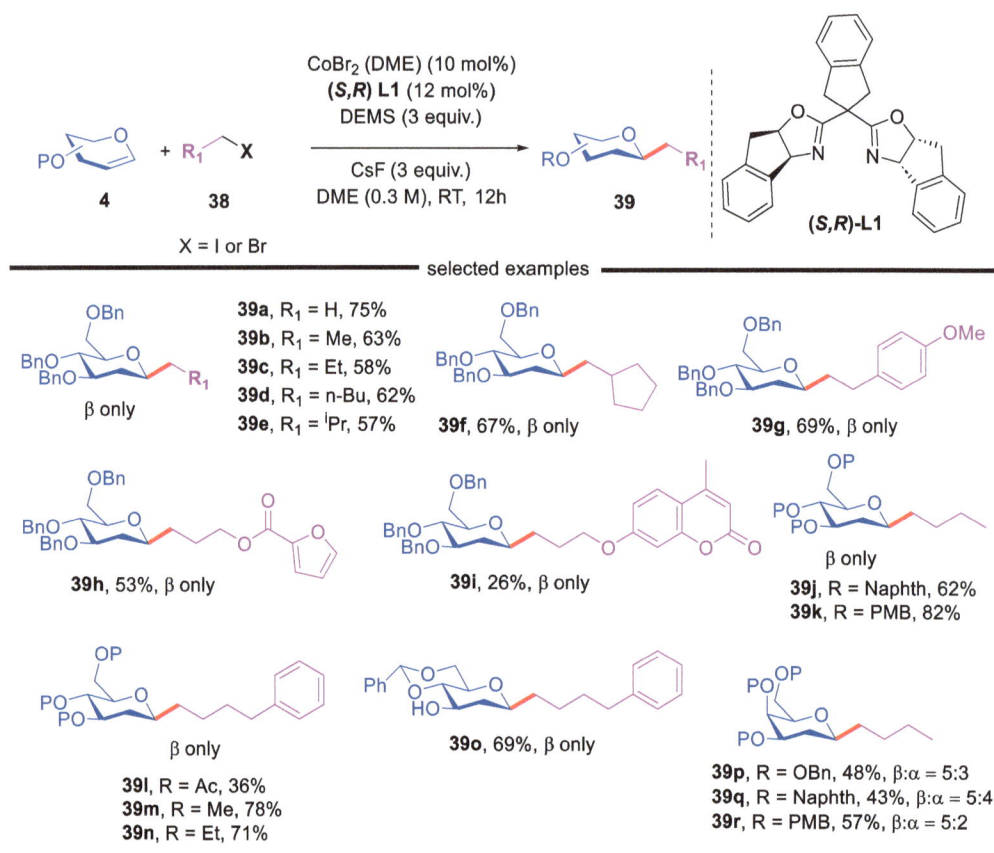

Scheme 17. Co-catalyzed synthesis of 2-deoxy-*C*-glycosides.

A proposed reaction pathway for the synthesis of 2-deoxy-*C*-glycosides **39** is depicted in (Scheme **18**). Initially, the catalyst $L_nCo^{II}X_2$ (**I**) generates the metal hydride $L_nCo^{II}X$-H (**II**) upon the reaction with silane. The intermediate **II**, on reaction with glycal **4,** gives the intermediate **III**, which can capture the alkyl radical quickly, resulting in the formation of Co^{III} species **IV**. Finally, the intermediate **IV** undergoes reductive elimination to produce the target compound **39** and Co(I) species **V**, which regenerates the initial catalyst $L_nCo^{II}X_2$ (**I**) and forms an alkyl radical by reacting with an alkyl halide.

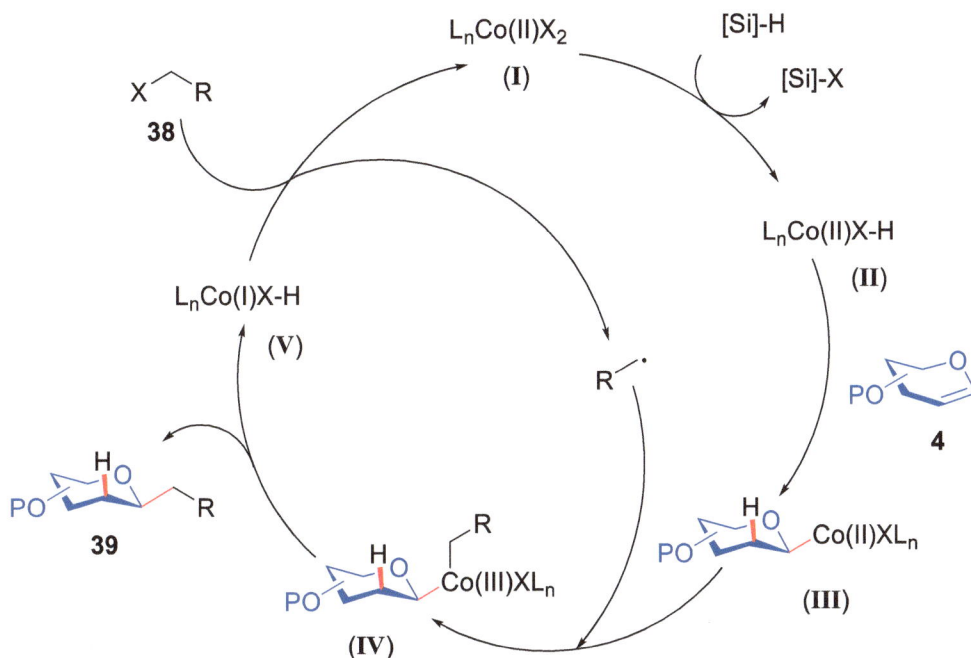

Scheme 18. Plausible mechanism for the co-catalyzed synthesis of 2-deoxy-*C*-glycosides.

Liu and their coworkers disclosed a ligand-controlled stereoselective *C*-alkyl glycosylation reaction to access 2-deoxy-β-*C*-alkyl glycosides **40** from glycals **4** and alkyl halides **39**, as illustrated in (Scheme **19**) [28]. Different benzyl-protected and acetyl-protected glycals, on reaction with their respective alkyl halides, afforded the desired products **41a-d** in good yields. However, tri-*O*-methyl-D-glucal particularly delivered various 2-deoxy-β-*C*-alkyl glycosides **41e-j** when reacted with the respective alkyl halides. The authors demonstrated that the method features broad substrate scope and excellent diastereoselectivity under mild conditions.

Mu *et al.* disclosed Co-catalyzed C3-glycosylation of indoles **41** using unfunctionalized glycals **4** for the production of 3-indolyl-*C*-deoxyglycosides **42**, as depicted in (Scheme **20**) [29]. Various indoles bearing different functional groups at the C-4, C-5, and C-6 positions, when subjected to the reaction with benzylated galactal, afforded the corresponding products (**42a-h**) in moderate yields. Furthermore, several protected glycal donors derived from D-glucose and D-galactose, on reaction with different indoles, provided the corresponding products **42n-o** and **42i-k**, respectively. However, the benzylated L-rhamnal delivered the corresponding product **42m** in good yield.

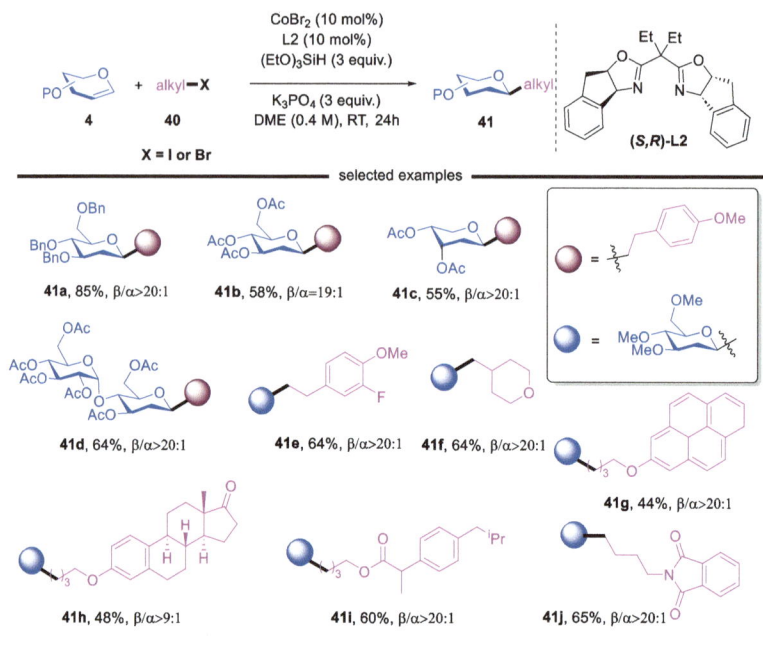

Scheme 19. Co-catalyzed synthesis of 2-deoxy-β-*C*-glycosides from glycals and alkyl halides.

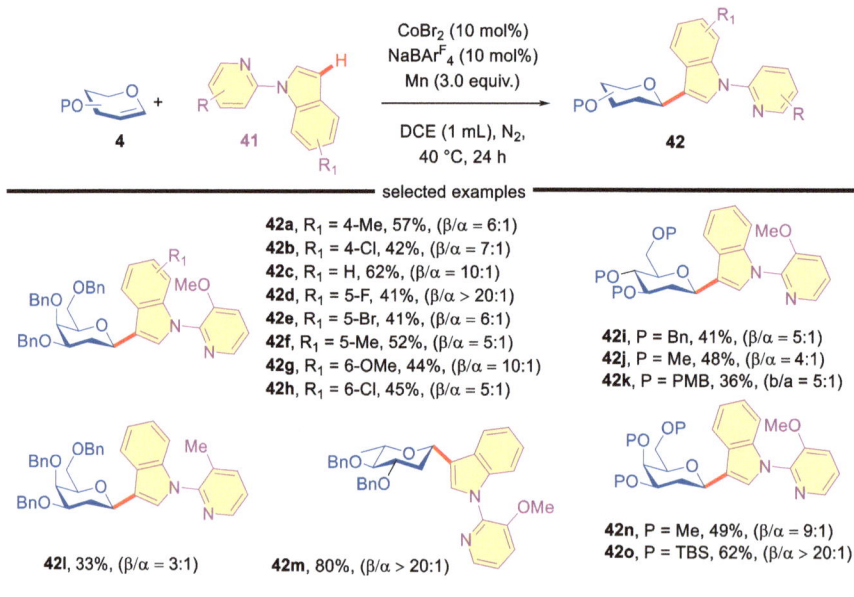

Scheme 20. Co-catalyzed synthesis of 3-indolyl-*C*-deoxyglycosides.

Nickel-catalyzed C-glycosylation

Li and co-workers have developed a novel technique for the production of 2-deoxy-α-*C*-glycosides **44** using a Ni-catalyzed C(sp³)-C(sp³) stereoselective coupling reaction of commercially available glycals **4** with alkyl halides **43**, as described in (Scheme **21**) [30]. A broad range of protected glycals were found to be tolerated for their reaction with different alkyl halides, affording the respective *C*-alkyl glycosides **44a-e** in good yields ranging from 51-86%. Furthermore, the reaction of benzylated glucal with various alkyl halides delivered the corresponding α-selective products **44f-n** in moderate to good yields ranging from 35% to 86%.

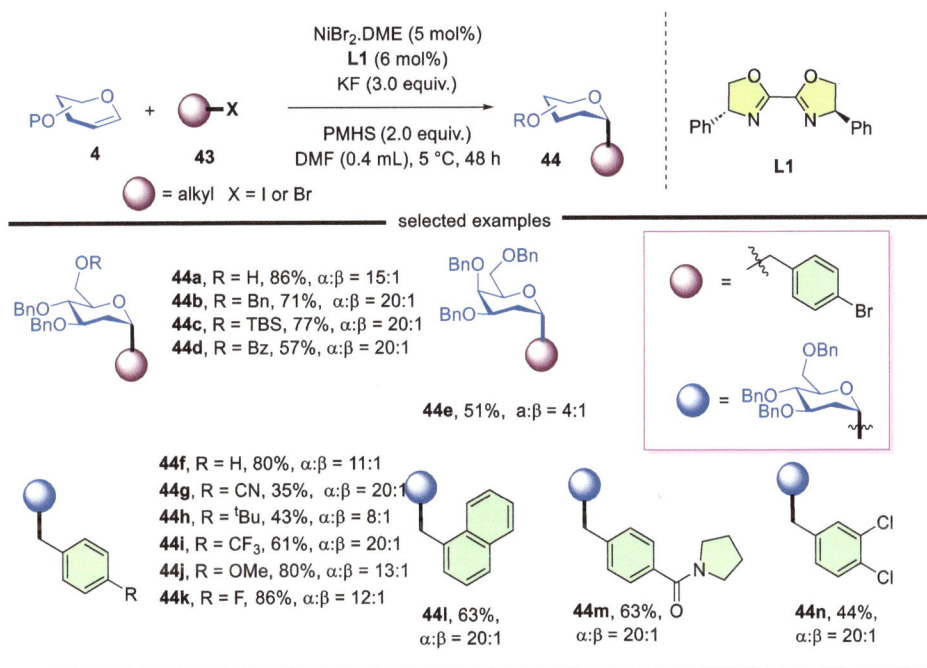

Scheme 21. Ni-catalyzed synthesis of 2-deoxy-α-*C*-glycosides.

Wang *et al.* described a Ni-catalyzed reductive decarboxylative/deaminative glycosylation approach of activated aliphatic acids/amines towards the preparation of alkyl *C*-glycosides [31]. In this study, TIPS-protected 1-iodoglycal **45** underwent the coupling reaction with several redox-active esters (RAE) **46** and Katritzky's *N*-alkylpyridinium salts **47** to afford alkyl *C*-glycosides **48-49** in good yields (Scheme **22**). The decarboxylative glycosylation was effective for several aliphatic RAE comprising different heterocyclic alkyl groups, affording the corresponding alkyl *C*-glycosides **44a-d** in moderate to excellent yields. However, the indomethacin-derived RAE delivered the product **48e** in 81% yield. Moreover, a variety of Katritzky's salts were coupled smoothly with **45** to form deaminated *C*-glycosides **49a-d** in 43-73% yields.

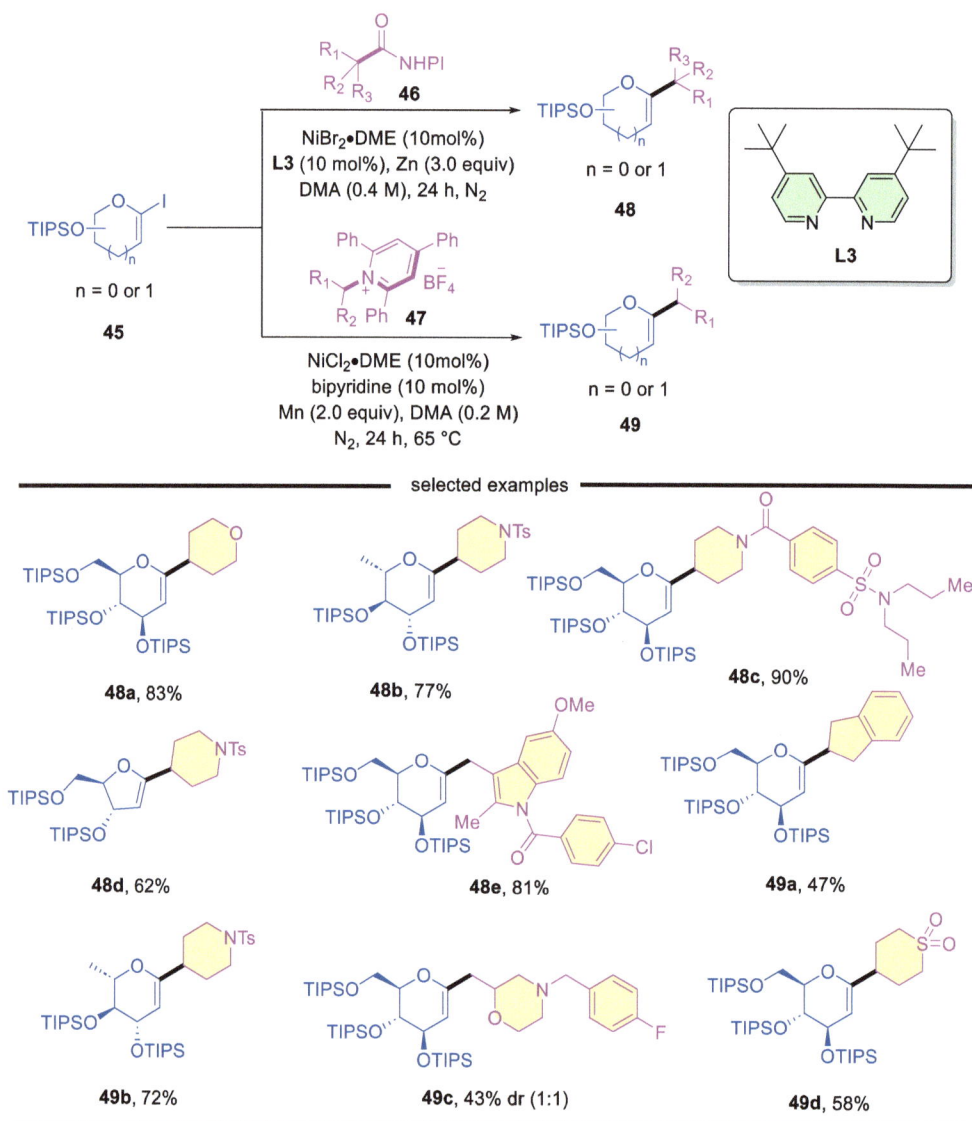

Scheme 22. Ni-catalyzed decarboxylative/deaminative glycosylation reaction.

A plausible reaction mechanism for the Ni(II)-catalyzed reductive decarboxylative/deaminative glycosylation is depicted in (Scheme **23**). The reaction begins with the reduction of the Ni(II) species by Zn or Mn, forming the active Ni(0) catalyst **I**, which then undergoes oxidative addition with 1-iodoglycal **45** to give the glycal-Ni^{II}(L) intermediate **II**. The Ni-complex **II** is then reduced by the reductant to form intermediate **III**, which undergoes a single-electron transfer (SET) with *N*-hydroxyphthalimide (NHPI) ester **46** or Katritzky's *N*-alkylpyridinium salts **47** to trigger decarboxylative/deaminative fragmentation to deliver the radical intermediate **V**, which then combines with the oxidized Ni-complex **IV** to form NiIII species **VI**. Finally, the intermediate **VI** undergoes reductive elimination to afford the desired product **4**.

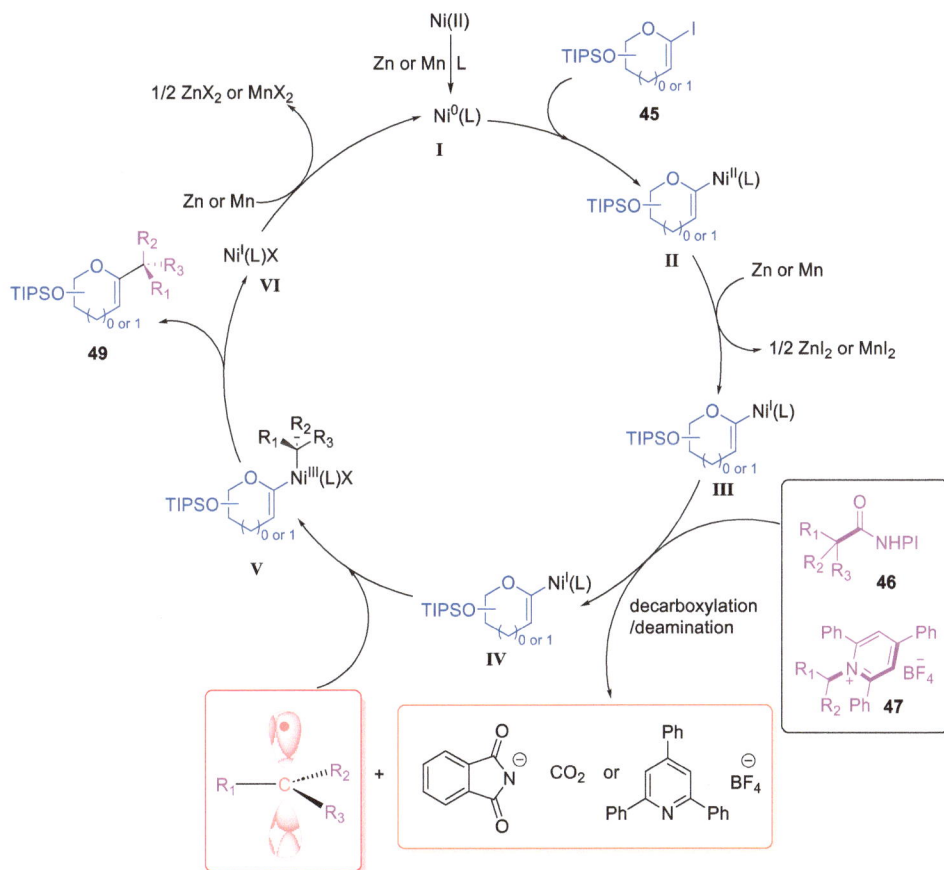

Scheme 23. Mechanistic pathway for the Ni-catalyzed decarboxylative/deaminative glycosylation reaction.

Ferry and their co-workers disclosed a novel approach for the preparation of 1-alkynylglycosides **52** derived from amidoquinoline (AQ) directed Ni-catalyzed pseudo-anomeric C-H alkynylation of C2-amidoglycal **50** with alkynyl bromides **51** under mild conditions, as illustrated in (Scheme **24**) [32]. The authors revealed that various alkynyl bromides containing silyl, cyclic propane, and TIPS substituents were smoothly reacted with different glycals, affording the corresponding *C*-glycosides **52a-c** & **52f**, **52d**, and **52e**, respectively. Furthermore, the late-stage functionalization of 1-alkynylglycosides featured through a click reaction to afford various *C*-glycoconjugates **52g-h** in excellent yields.

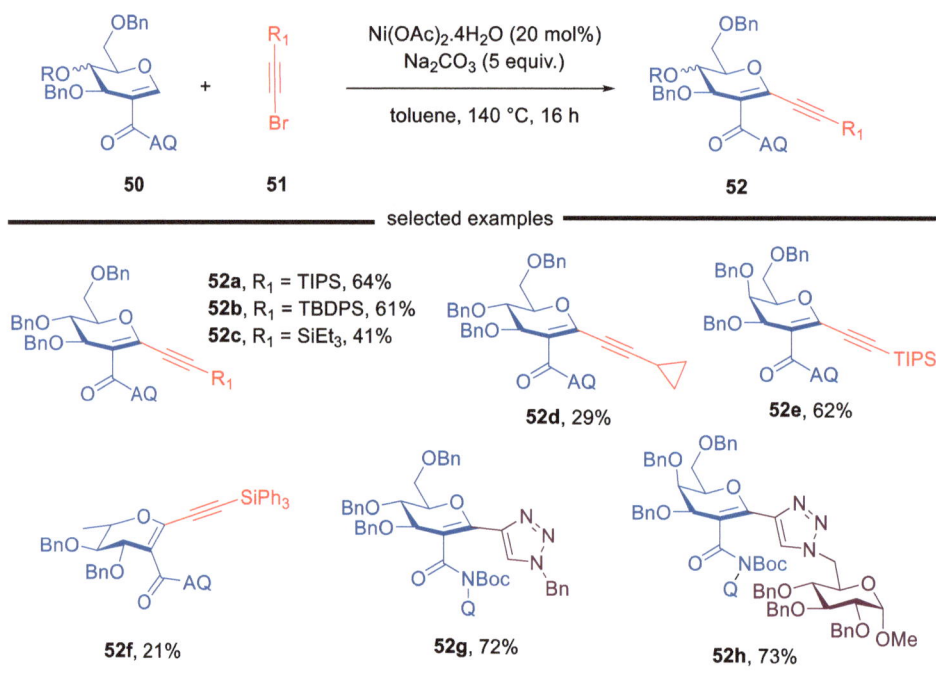

Scheme 24. Ni-catalyzed pseudo-anomeric C-H alkynylation of C2-amidoglycal with alkynyl bromides.

Iridium-Catalyzed C-glycosylation

Zhu *et al.* developed a practical method for accessing 2-deoxy-*C*-aryl **55** and alkyl-**56** glycosides through Ir-catalyzed benzoxazole (BO) directed sp^2 and sp^3 C-H glycosylation with diverse glycals **4** obtained from D-glucose, D-galactose, and L-rhamnose, as depicted in (Scheme **25**) [33]. The authors disclosed that the method was acquiescent for both the C-H glycosylation of indoles **53** and the secondary alkyl amines **54**. The differently substituted indoles containing electron-donating groups (EDGs) produced the products **55a-b** and **55d-e** in good to excellent yields with high β-diastereoselectivity. However, the indole bearing an electron-withdrawing group (EWG) reduced the β-diastereoselectivity, affording the product **55c** in 87% yields. Moreover, substrate **54** reacted with glycals **4** to give the desired 2-deoxy-*C*-alkyl glycosides **56a-d** in 52-76% yields. Furthermore, the authors disclosed that the benzoxazole directing group could be effectively removed by the treatment with KOH or LiAlH$_4$, leading to various value-added products.

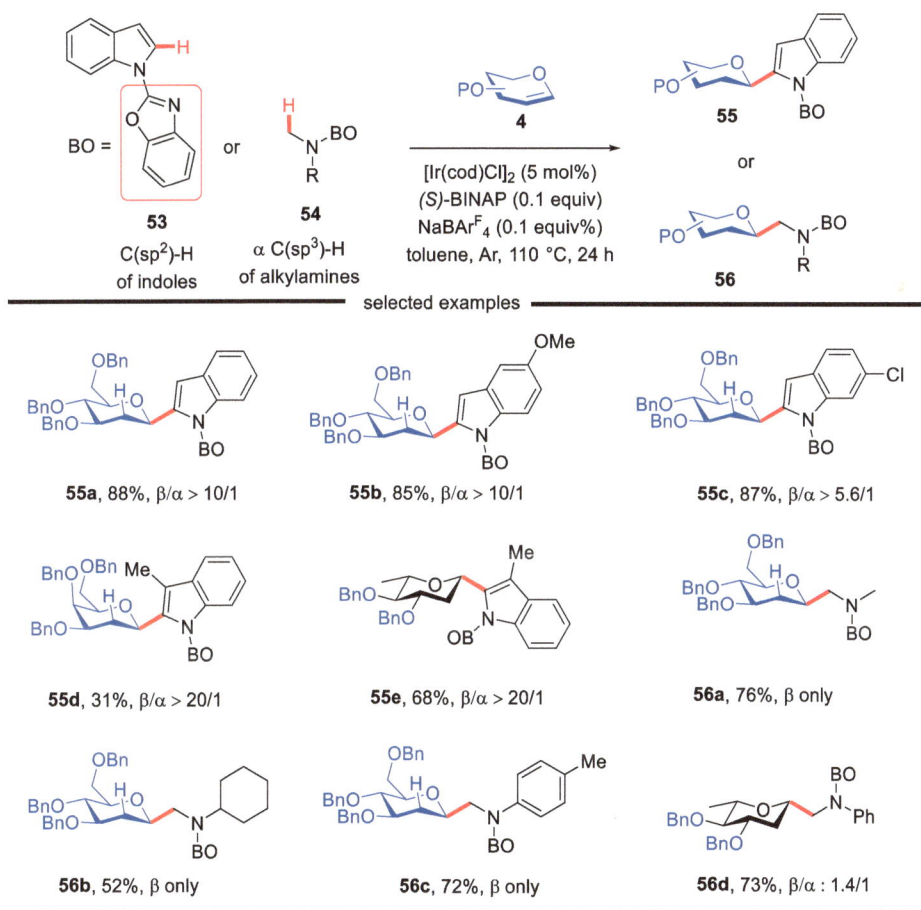

Scheme 25. Ir-catalyzed benzoxazole (BO)-directed sp^2/ sp^3 C-H glycosylation of indoles and alkyl amines.

Yu *et al.* demonstrated a competent method for the regio- and stereoselective production of 2-indolyl-*C*-deoxyglycosides **57** through Ir(I)-catalyzed and pyridine-directed C-H functionalization of various indoles **56** with different protected glycals **4**, as demonstrated in (Scheme **26**) [34]. When the benzylated D-glucal was subjected to the C-H functionalization reaction with different indole substrates **56**, it delivered 2-indolyl-*C*-deoxyglycosides **57a-e** in moderate to excellent yields. The glycal donors derived from glucose, galactose, and rhamnose bearing the different protecting groups reacted smoothly with different indoles to access the desired products **57h-j**, **57k-m**, and **57n-o**, respectively.

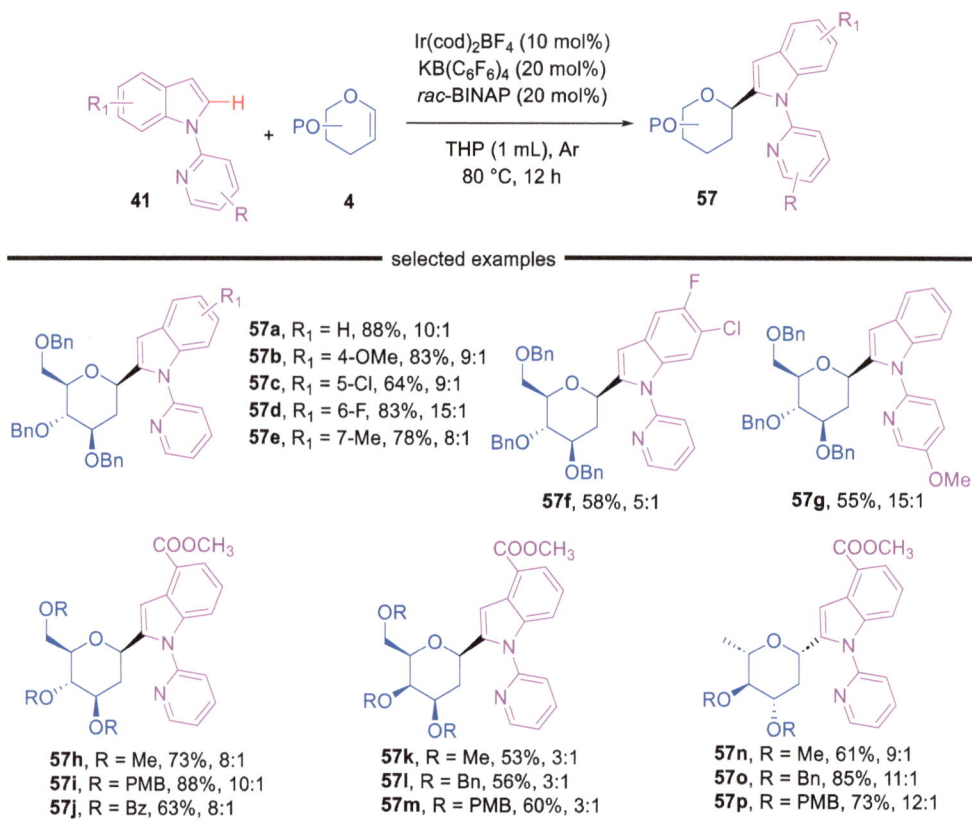

Scheme 26. Ir(I)-catalyzed and pyridine group-directed C-H functionalization of indoles with glycals.

Other Types of *C*-glycosylation

Hu and coworkers have demonstrated a practical approach for synthesizing 1,2-*trans*-*C*-alkyl glycosides **60** from 2-deoxy-2-amino sugars **58** (2DASs) involving glycosyl radicals using photo-redox catalysis, as depicted in (Scheme **27**) [35]. An imidate derived from D-glucal, on reaction with differently substituted styrenes, delivered the corresponding *C*-alkyl glycosides **60a-e** in 43-

92% yields. Several glycal donors, including cyclic ketal/acetal groups containing different protecting groups such as PMB, di-*tert*-butyl silylene, and benzylidene, were well compatible under the standard conditions, affording the corresponding products **60f-h** in good yields. Moreover, the imidates derived from D-allal and D-galactal produced β-selective products **60i** (81%) and **60j** (76%) in an α-selective manner, respectively.

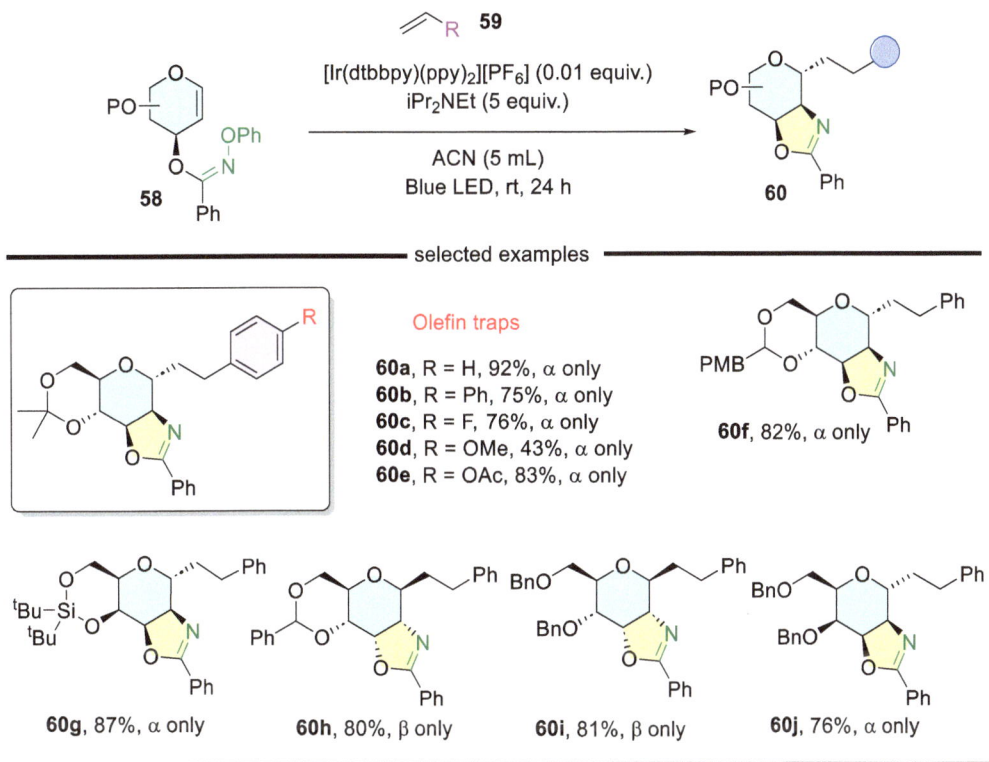

Scheme 27. Ir-catalyzed synthesis of *C*-glycosides derived from 2-deoxy-2-amino sugars (2DASs).

The research group of Sagar has developed a metal-free approach for synthesizing chirally enriched pyrazolylpyrimidinone-based glycohybrids (**63-64**) from various 2-hydrazineylpyrimidin-4(3*H*)-ones **61** and benzyl protected 2-*C*-formyl glycals **62** under microwave irradiation in ethanol, as illustrated in (Scheme **28**) [36]. The authors applied the method to different 2-*C*-formyl glycals derived from benzylated D-glucal and D-galactal to access the corresponding glycohybrids **63a-c** and **64a-c**, respectively. Moreover, the authors disclosed that the protocol exhibits operational simplicity and broad substrate scope under mild reaction conditions.

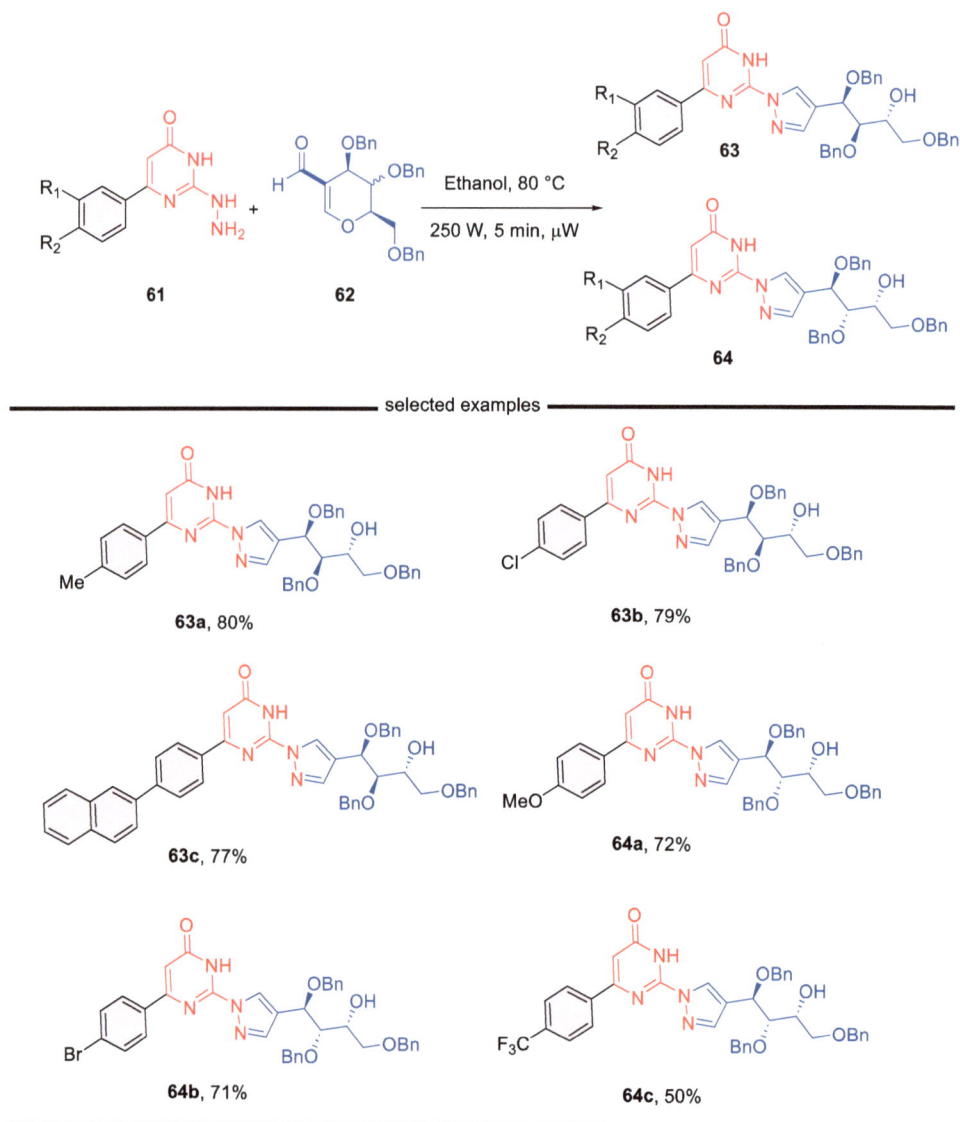

Scheme 28. Metal-free synthesis of chirally enriched pyrazolylpyrimidinone-based glycohybrids.

Sagar and co-workers have disclosed a catalyst-free technique for the creation of chirally enriched imidazo [1,2-*a*]pyrimidinone glycohybrids **67-68** using a base-induced annulation of α-iodo-pyranone **65** with 2-aminopyrimidinones **66**, as described in (Scheme **29**) [37]. A glucal-based iodo-pyranone, on reaction with differently substituted 2-aminopyrimidin-4(1*H*)-ones, afforded the respective products **67a-c** in good yields. However, galactal-based iodo-pyranone, on reaction with differently substituted 2-aminopyrimidin-4(1*H*)-ones, delivered the respective products **68a-c** in good yields.

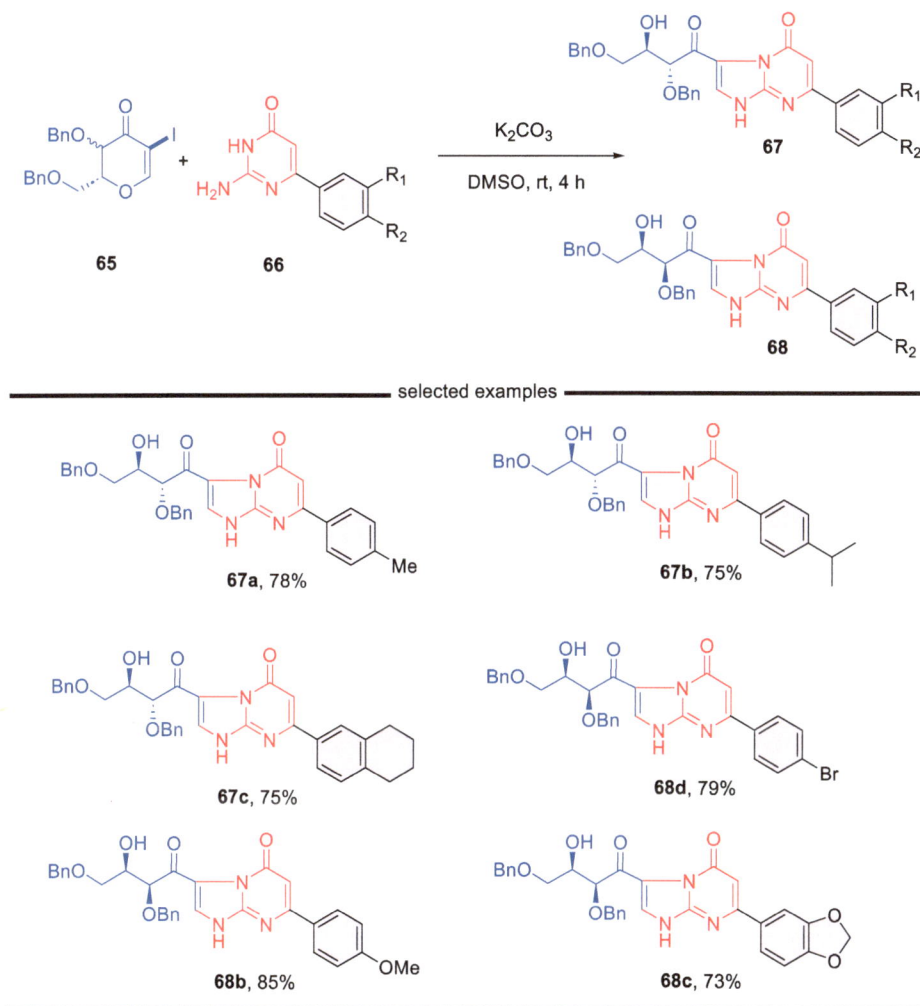

selected examples

Scheme 29. Catalyst-free synthesis of chirally enriched imidazo [1,2-a]pyrimidinone glycohybrids.

Arora *et al.* reported a resourceful protocol for the construction of sugar-derived pyrano [3,2-c]quinolones **70** employing 1-*C*-formyl glycal **62'** and 4-hydroxy quinolones **69** followed by 1,2-annulation, as shown in (Scheme **30**) [38]. As evident from (Scheme **30**), a number of electron-donating and electron-withdrawing substituents such as methyl, isopropyl, bromo, fluoro, and trifluoromethoxy on the 4-hydroxy quinolone moiety were tolerated well in the reaction, affording the respective products **70a-f** in good yields. The authors disclosed that the molecular docking of the synthesized natural product analogs validates their binding modes within the active site of type II topoisomerase.

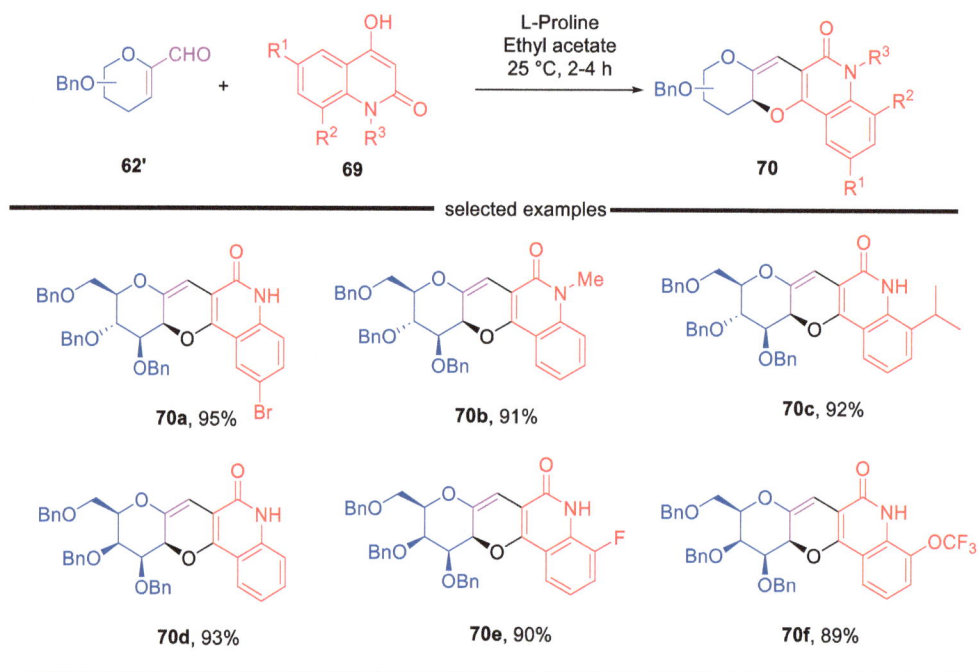

Scheme 30. Synthesis of sugar-derived pyrano [3,2-c]quinolones.

Guo *et al.* disclosed β-stereoselective glycosylation of indoles **72** with glycals **71** by employing a phosphonoselenide catalyst (**Cat B**) to access indolyl *C*-glycosides **73**, as depicted in (Scheme **31**) [39]. A plethora of electron-donating and electron-withdrawing substituents on the indole moiety were well tolerated in C3-glycosylation, affording the corresponding products **73a-q** in good yields. The strength of the protocol was illustrated by its amenability for reaction at both the indolyl *C*- and *N*-reactive sites. Additionally, the authors disclosed that the computational studies and NMR spectroscopy reveals that the chalcogenic and aromatic components of the catalyst can be collectively exploited to foster conformational distortion of glycal away from the usual half-chair to the boat conformation, releasing the convex β-face for the nucleophilic attack exclusively.

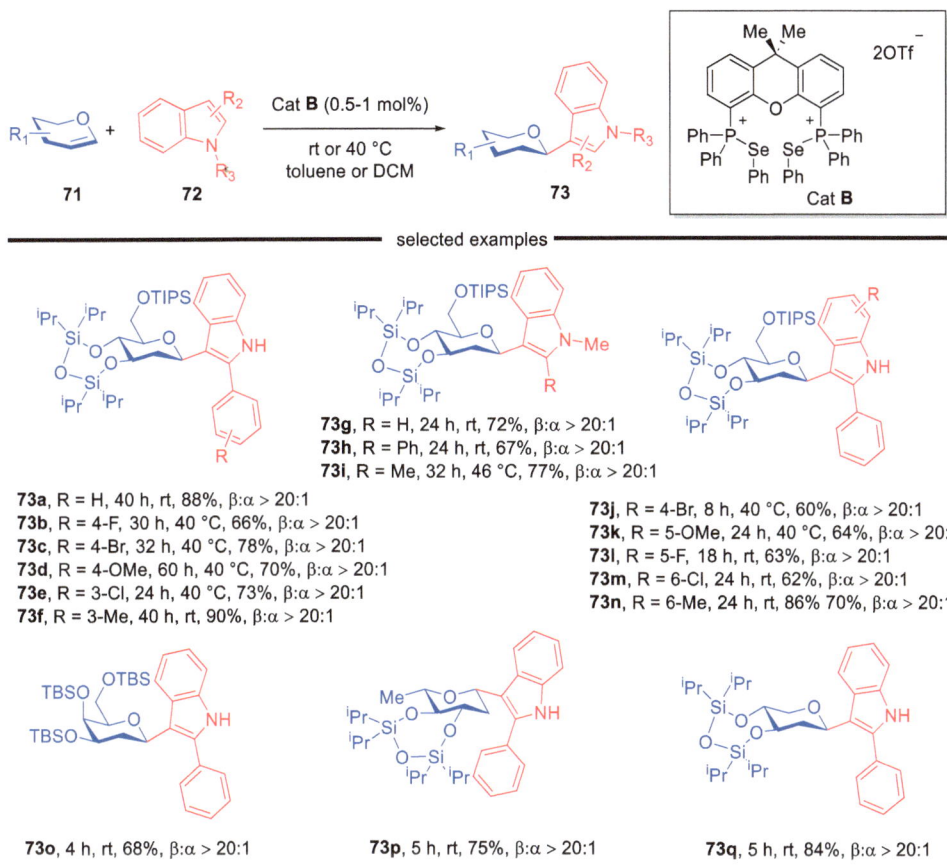

Scheme 31. Metal-free β-stereoselective glycosylation of indoles with glycals.

CONCLUSION

The chemical *C*-glycosylation reactions are always at the core of carbohydrate chemistry owing to their extensive applications in medicinal chemistry and drug development. Several novel tactics have been devised for one-pot synthesis of value-added *C*-glycosides using glycals and their derivatives. Ultimately, this chapter highlights the recent synthetic strategies employed for synthesizing privileged *C*-glycosides by adopting metal-catalyzed and metal-free protocols under mild conditions.

REFERENCES

[1]   Hussain, N.; Hussain, A. Advances in Pd-catalyzed C–C bond formation in carbohydrates and their applications in the synthesis of natural products and medicinally relevant molecules. *RSC Advances,* **2021**, *11*(54), 34369-34391.
[http://dx.doi.org/10.1039/D1RA06351K] [PMID: 35497292]

[2]   Gou, X.Y.; Zhu, X.Y.; Zhang, B.S.; Liang, Y.M. Synthesis of c-aryl glycosides by c–h functionalization. *Chemistry,* **2023**, *29*(32), e202203351.
[http://dx.doi.org/10.1002/chem.202203351] [PMID: 36943394]

[3]   Parida, S.P.; Das, T.; Ahemad, M.A.; Pati, T.; Mohapatra, S.; Nayak, S. Recent advances on synthesis of *C*-glycosides. *Carbohydr. Res.,* **2023**, *530*, 108856.
[http://dx.doi.org/10.1016/j.carres.2023.108856] [PMID: 37315353]

[4]   Singh, Y.; Geringer, S.A.; Demchenko, A.V. Synthesis and glycosidation of anomeric halides: evolution from early studies to modern methods of the 21st century. *Chem. Rev.,* **2022**, *122*(13), 11701-11758.
[http://dx.doi.org/10.1021/acs.chemrev.2c00029] [PMID: 35675037]

[5]   Kitamura, K.; Ando, Y.; Matsumoto, T.; Suzuki, K. Total synthesis of aryl *c* -glycoside natural products: strategies and tactics. *Chem. Rev.,* **2018**, *118*(4), 1495-1598.
[http://dx.doi.org/10.1021/acs.chemrev.7b00380] [PMID: 29281269]

[6]   Bokor, É.; Kun, S.; Goyard, D.; Tóth, M.; Praly, J.P.; Vidal, S.; Somsák, L. *C* -glycopyranosyl arenes and hetarenes: synthetic methods and bioactivity focused on antidiabetic potential. *Chem. Rev.,* **2017**, *117*(3), 1687-1764.
[http://dx.doi.org/10.1021/acs.chemrev.6b00475] [PMID: 28121130]

[7]   Wang, X.; Zhang, L.; Byrne, D.; Nummy, L.; Weber, D.; Krishnamurthy, D.; Yee, N.; Senanayake, C.H. Efficient synthesis of Empagliflozin, an inhibitor of SGLT-2, utilizing an AlCl$_3$-promoted silane reduction of a β-glycopyranoside. *Org. Lett.,* **2014**, *16*(16), 4090-4093.
[http://dx.doi.org/10.1021/ol501755h] [PMID: 25061799]

[8]   Chao, E.C.; Henry, R.R. SGLT2 inhibition — a novel strategy for diabetes treatment. *Nat. Rev. Drug Discov.,* **2010**, *9*(7), 551-559.
[http://dx.doi.org/10.1038/nrd3180] [PMID: 20508640]

[9]   Wei, A.; Boy, K.M.; Kishi, Y. Biological evaluation of rationally modified analogs of the h-type ii blood group trisaccharide. a correlation between solution conformation and binding affinity. *J. Am. Chem. Soc.,* **1995**, *117*(37), 9432-9436.
[http://dx.doi.org/10.1021/ja00142a008]

[10]   Franck, R.W.; Tsuji, M. α-c-galactosylceramides: synthesis and immunology. *Acc. Chem. Res.,* **2006**, *39*(10), 692-701.
[http://dx.doi.org/10.1021/ar050006z] [PMID: 17042469]

[11]   Yang, G.; Schmieg, J.; Tsuji, M.; Franck, R.W. The *C*-glycoside analogue of the immunostimulant alpha-galactosylceramide (KRN7000): synthesis and striking enhancement of activity. *Angew. Chem. Int. Ed.,* **2004**, *43*(29), 3818-3822.
[http://dx.doi.org/10.1002/anie.200454215] [PMID: 15258945]

[12]   Liu, C.-F. Recent Advances on Natural Aryl-C-Glycoside Scaffolds: Structure, Bioactivities, and Synthesis—A Comprehensive Review. Molecules 2022; 27 (21), 7439.
[http://dx.doi.org/10.3390/molecules27217439]

[13]   Kinfe, H.H. Versatility of glycals in synthetic organic chemistry: coupling reactions, diversity oriented synthesis and natural product synthesis. *Org. Biomol. Chem.,* **2019**, *17*(17), 4153-4182.
[http://dx.doi.org/10.1039/C9OB00343F] [PMID: 30893410]

[14]   Yao, Y.; Xiong, C.P.; Zhong, Y.L.; Bian, G.W.; Huang, N.Y.; Wang, L.; Zou, K. Intramolecular and ferrier rearrangement strategy for the construction of c1- *β* -d -xylopyranosides: synthesis, mechanism and biological activity study. *Adv. Synth. Catal.,* **2019**, *361*(5), 1012-1017.
[http://dx.doi.org/10.1002/adsc.201801423]

[15]   Addanki, R.B.; Halder, S.; Kancharla, P.K. TfO$^-$ ···H–O–H interaction-assisted generation of a silicon cation from allylsilanes: access to phenylallyl ferrier glycosides from glycals. *Org. Lett.,* **2022**, *24*(7), 1465-1470.

[http://dx.doi.org/10.1021/acs.orglett.2c00062] [PMID: 35142527]

[16]   Wei, X.; Zeng, M.; Li, Y.; Wang, D.; Wang, J.; Liu, H. Palladium(II)-catalyzed heck coupling: direct stereoselective synthesis of *c* -aryl glycosides from nonactivated glycals and thianthrenium salts. *Org. Lett.,* **2024**, *26*(12), 2473-2477.
[http://dx.doi.org/10.1021/acs.orglett.4c00654] [PMID: 38498594]

[17]   Pal, K.B.; Lee, J.; Das, M.; Liu, X.W. Palladium(ii)-catalyzed stereoselective synthesis of *C* -glycosides from glycals with diaryliodonium salts. *Org. Biomol. Chem.,* **2020**, *18*(12), 2242-2251.
[http://dx.doi.org/10.1039/D0OB00247J] [PMID: 32159571]

[18]   Sharma, M.K.; Tiwari, B.; Hussain, N. Pd-catalyzed stereoselective synthesis of chromone *C* -glycosides. *Chem. Commun. (Camb.),* **2024**, *60*(36), 4838-4841.
[http://dx.doi.org/10.1039/D4CC00486H] [PMID: 38619439]

[19]   Singh, A.K.; Kanaujiya, V.K.; Tiwari, V.; Sabiah, S.; Kandasamy, J. Development of routes for the stereoselective preparation of β-aryl- *c* -glycosides *via c* -1 aryl enones. *Org. Lett.,* **2020**, *22*(19), 7650-7655.
[http://dx.doi.org/10.1021/acs.orglett.0c02843] [PMID: 32941050]

[20]   Dubey, A.; Singh Chauhan, N.; Azeem, Z.; Kumar Mandal, P. Directed palladium-catalyzed *pseudo* -anomeric c−h functionalization of glycal-type substrates: access to unsymmetrical *gem* -diarylmethyl *c* -glycosides. *Adv. Synth. Catal.,* **2023**, *365*(6), 820-825.
[http://dx.doi.org/10.1002/adsc.202201343]

[21]   Xiao, X.; Han, P.; Wan, J.P.; Liu, J. Stereoselective synthesis of indolyl- *c* -glycosides enabled by sequential aminopalladation and heck glycosylation of 2-alkynylanilines with glycals. *Org. Lett.,* **2023**, *25*(39), 7170-7175.
[http://dx.doi.org/10.1021/acs.orglett.3c02688] [PMID: 37756216]

[22]   Liu, J.; Xiao, X.; Han, P.; Zhou, H.; Yin, Q.S.; Sun, J.S. Palladium-catalyzed *C* -glycosylation and annulation of *o* -alkynylanilines with 1-iodoglycals: convenient access to 3-indolyl- *C* -glycosides. *Org. Biomol. Chem.,* **2020**, *18*(43), 8834-8838.
[http://dx.doi.org/10.1039/D0OB01812K] [PMID: 33103171]

[23]   Wu, J.; Kopp, A.; Ackermann, L. Synthesis of *c* -oligosaccharides through versatile c(sp^3)−h glycosylation of glycosides. *Angew. Chem. Int. Ed.,* **2022**, *61*(11), e202114993.
[http://dx.doi.org/10.1002/anie.202114993] [PMID: 35015329]

[24]   Ding, W.Y.; Zhao, H.W.; Cheng, J.K.; Lu, Z.; Xiang, S.H.; Tan, B. β-*C*-glycosylation with 2-oxindole acceptors *via* palladium-catalyzed decarboxylative reactions. *Org. Lett.,* **2022**, *24*(38), 7031-7036.
[http://dx.doi.org/10.1021/acs.orglett.2c02881] [PMID: 36129413]

[25]   Vaňková, K.; Rahm, M.; Choutka, J.; Pohl, R.; Parkan, K. Facile approach to *c* -glucosides by using a protecting-group-free hiyama cross-coupling reaction: high-yielding dapagliflozin synthesis. *Chemistry,* **2021**, *27*(41), 10583-10588.
[http://dx.doi.org/10.1002/chem.202101052] [PMID: 34048112]

[26]   Liu, Y.; Wang, Y.; Dai, W.; Huang, W.; Li, Y.; Liu, H.; Palladium-Catalysed, C. Palladium-catalysed c(sp^3)−h glycosylation for the synthesis of c-alkyl glycoamino acids. *Angew. Chem. Int. Ed.,* **2020**, *59*(9), 3491-3494.
[http://dx.doi.org/10.1002/anie.201914184] [PMID: 31901005]

[27]   Zeng, M.; Yu, C.; Wang, Y.; Wang, J.; Wang, J.; Liu, H. Cobalt(II)-catalyzed c(sp^3)−c(sp^3) coupling for the direct stereoselective synthesis of 2-deoxy- *c* -glycosides from glycals. *Angew. Chem. Int. Ed.,* **2023**, *62*(22), e202300424.
[http://dx.doi.org/10.1002/anie.202300424] [PMID: 36929518]

[28]   Liu, B.; Liu, D.; Rong, X.; Lu, X.; Fu, Y.; Liu, Q. Ligand-controlled stereoselective synthesis of 2-deoxy-β-c-glycosides by cobalt catalysis. *Angew. Chem. Int. Ed.,* **2023**, *62*(22), e202218544.
[http://dx.doi.org/10.1002/anie.202218544] [PMID: 36929313]

[29]   Mu, Q.Q.; Guo, A.X.; Cai, X.; Qin, Y.Y.; Liu, X.L.; Ye, F.Z.; Yang, H.J.; Xiao, X.; Liu, X.W. Cobalt's dual role in promoting c3-glycosylation of indoles: unraveling mechanistic insights. *Org. Lett.,* **2023**, *25*(38), 7040-7045.
        [http://dx.doi.org/10.1021/acs.orglett.3c02624] [PMID: 37721454]

[30]   Shi, H.; Yin, G.; Lu, X.; Li, Y. Stereoselective synthesis of 2-deoxy-α-*C*-glycosides from glycals. *Chin. Chem. Lett.,* **2024**, *35*(12), 109674.
        [http://dx.doi.org/10.1016/j.cclet.2024.109674]

[31]   Liu, X.G.; Yang, Q.; Liu, D.Y.; Liu, J.; Tan, D.H.; Ruan, Y.J.; Wang, P.F.; Wang, X.L.; Wang, H. Nickel-catalyzed reductive decarboxylative/deaminative glycosylation of activated aliphatic acids and primary amines. *Org. Lett.,* **2023**, *25*(27), 5022-5026.
        [http://dx.doi.org/10.1021/acs.orglett.3c01691] [PMID: 37395740]

[32]   de Robichon, M.; Branquet, D.; Uziel, J.; Lubin-Germain, N.; Ferry, A. Directed nickel-catalyzed *pseudo* -anomeric c−h alkynylation of glycals as an approach towards *c* -glycoconjugate synthesis. *Adv. Synth. Catal.,* **2021**, *363*(22), 5138-5148.
        [http://dx.doi.org/10.1002/adsc.202100823]

[33]   Zhu, W.; Sun, Q.; Chang, H.; Zhang, H.X.; Wang, Q.; Chen, G.; He, G. Synthesis of2-deoxy- *c*-glycosides *via* iridium-catalyzed sp^2 and sp^3 c—h glycosylation with unfunctionalized glycals. *Chin. J. Chem.,* **2022**, *40*(5), 571-576.
        [http://dx.doi.org/10.1002/cjoc.202100658]

[34]   Yu, C.; Liu, Y.; Xie, X.; Hu, S.; Zhang, S.; Zeng, M.; Zhang, D.; Wang, J.; Liu, H. Ir(I)-catalyzed c−h glycosylation for synthesis of 2-indolyl- *c* -deoxyglycosides. *Adv. Synth. Catal.,* **2021**, *363*(21), 4926-4931.
        [http://dx.doi.org/10.1002/adsc.202100855]

[35]   Shi, W.Z.; Li, H.; Mu, G.C.; Lu, J.L.; Tu, Y.H.; Hu, X.G. 1,2- *trans* -stereoselective synthesis of *c* -glycosides of 2-deoxy-2-amino-sugars involving glycosyl radicals. *Org. Lett.,* **2021**, *23*(7), 2659-2663.
        [http://dx.doi.org/10.1021/acs.orglett.1c00551] [PMID: 33733785]

[36]   Tiwari, G.; Mishra, V.K.; Khanna, A.; Tyagi, R.; Sagar, R. Synthesis of chirally enriched pyrazolylpyrimidinone-based glycohybrids *via* annulation of glycals with 2-hydrazineylpyrimidin- 4(3 *h*)-ones. *J. Org. Chem.,* **2024**, *89*(7), 5000-5009.
        [http://dx.doi.org/10.1021/acs.joc.4c00211] [PMID: 38471017]

[37]   Mishra, V.K.; Tiwari, G.; Khanna, A.; Yadav, Y.; Sagar, R. Base-induced annulation of glycal-derived α-iodopyranone with 2-aminopyrimidinones: access to chiral imidazopyrimidinones. *Eur. J. Org. Chem.,* **2024**, *27*(8), e202301301.
        [http://dx.doi.org/10.1002/ejoc.202301301]

[38]   Arora, A.; Kumar, S.; Kumar, S.; Singh, S.K.; Dua, A.; Singh, B.K. Natural product inspired diastereoselective synthesis of sugar-derived pyrano[3,2-c]quinolones and their *in-silico* studies. *Carbohydr. Res.,* **2024**, *539*, 109105.
        [http://dx.doi.org/10.1016/j.carres.2024.109105] [PMID: 38583285]

[39]   Guo, H.; Kirchhoff, J.L.; Strohmann, C.; Grabe, B.; Loh, C.C.J. Exploiting π and chalcogen interactions for the β-selective glycosylation of indoles through glycal conformational distortion. *Angew. Chem. Int. Ed.,* **2024**, *63*(7), e202316667.
        [http://dx.doi.org/10.1002/anie.202316667] [PMID: 38116860]

Exploring the C-2 Position of Glycals: Structural Insights and Synthetic Applications

Ram Pratap Pandey[1]**, Manish Kumar Sharma**[1]**, Anand Kumar Pandey**[1]**, Altaf Hussain**[2] **and Nazar Hussain**[1,*]

[1] *Department of Medicinal Chemistry, Institute of Medical Sciences, Banaras Hindu University, Varanasi-221005, India*

[2] *Government Degree College, Budhal, J&K Higher Education Department, Jammu and Kashmir-185233, India*

Abstract: The chemistry of glycals and their derivatives has emerged as a hot topic in carbohydrate chemistry owing to their incredible applications in biological and medicinal chemistry. Annulated and C-2-branched sugars derived from glycal moiety and its derivative have received immense attention. Herein, we have incorporated the current advancement in the synthesis of 2-C-branched sugars and annulated sugars derived from glycals, 2-haloglycals, and 2-nitroglycals using various synthetic strategies, including C-H activation, 1,2-annulation, and cyclopropanation.

Keywords: C-H activation, Cyclopropanation, Glycals, Glycosides, 1,2-Annulation, 2-Haloglycals.

INTRODUCTION

Glycals are a class of carbohydrates distinguished by a cyclic structure that includes an oxygen atom and an unsaturated alkene bond between the 1-C and 2-C positions of the pyranose or furanose ring. These compounds are crucial in modern organic synthesis and mimic structural elements found in biologically active molecules. The first glycal was discovered and synthesized by Fischer and Zach in 1913 [1]. The double bond within the ring enables various reactions, including addition, cycloaddition, substitution, cross-coupling, 1,2-annulation, rearrangements, and occasionally ring opening, as illustrated in Fig. (**1**) [2]. The oxygen atom in the ring enhances conjugation and promotes the trapping of nucleophiles at the C-1 position and electrophiles at the C-2 position.

---

* **Corresponding author Nazar Hussain:** Department of Medicinal Chemistry, Institute of Medical Sciences, Banaras Hindu University, Varanasi-221005, India; E-mail: nazar10@bhu.ac.in

Nazar Hussain & Atul Kumar (Eds.)

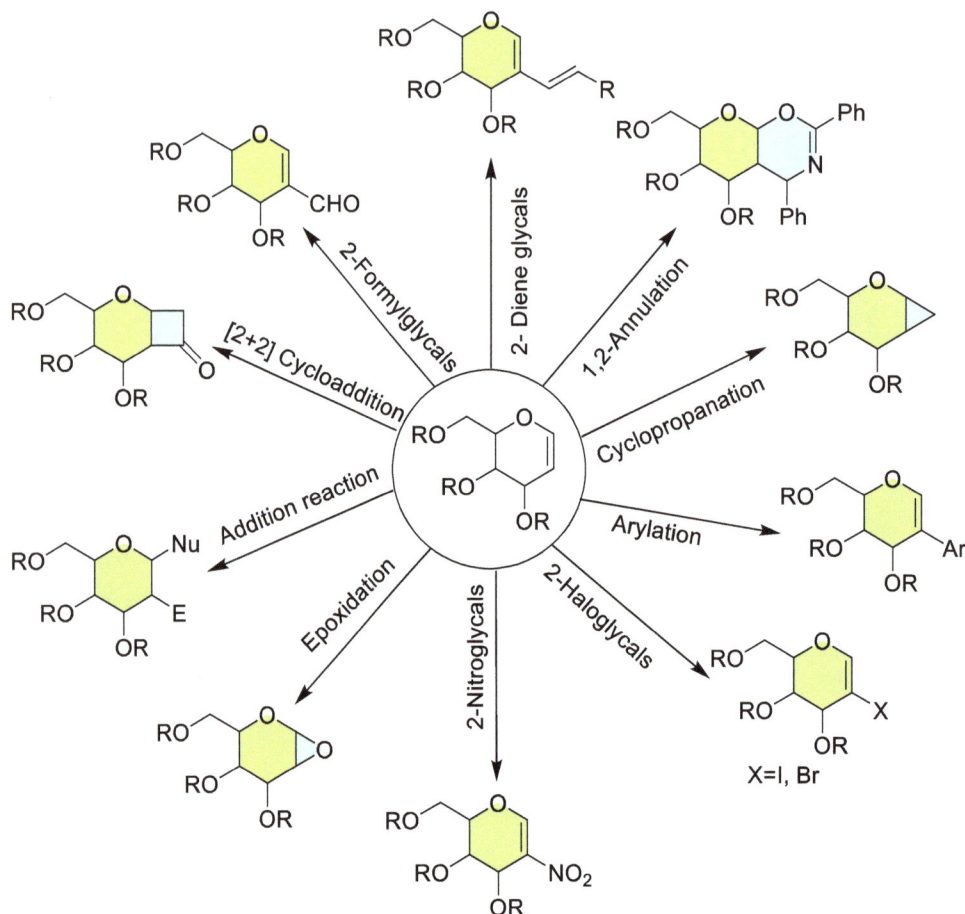

Fig. (1). Typical reactivity of glycals.

The most studied transformations of glycals involve a rearrangement known as Ferrier rearrangement, in which the double bond is usually shifted from the carbon atom 1 and 2 position to the ring's 2 and 3 positions, usually under acidic conditions. The C-2 position of glycal is nucleophilic, which allows various electrophilic attacks to generate 2-C-branched sugars [3]. The presence of a double bond inside the ring makes the glycals prone to epoxidation, hydroxylation, hydrogenation, ozonolysis, *etc.* [4]. As a result, the glycals can be further employed in different types of glycosylation reactions, 1,2-annulation reactions and C-2 branched synthesis [5, 6]. In the recent past, a large number of efforts have been made for reactions at the C-2 position of glycals. 2-C-branching and 1,2-annulation reactions are particularly appealing because they allow the manipulation of the anomeric center. Often, C-2-branched sugars can be further transformed into 1,2-annulated sugars, which serve as valuable precursors for synthesizing a variety of biologically relevant compounds. The development of these molecules aids in creating novel synthetic methods to assemble these small molecules into chiral building blocks for complex compounds.

Over the past decades, the reaction at the C2 position of glycal can offer a cluster of C2 functionalized products with enormous advantages. The C2 functionalized glycals are widely distributed in numerous natural molecules and biologically active motifs, as outlined in Fig. (**2**) [7].

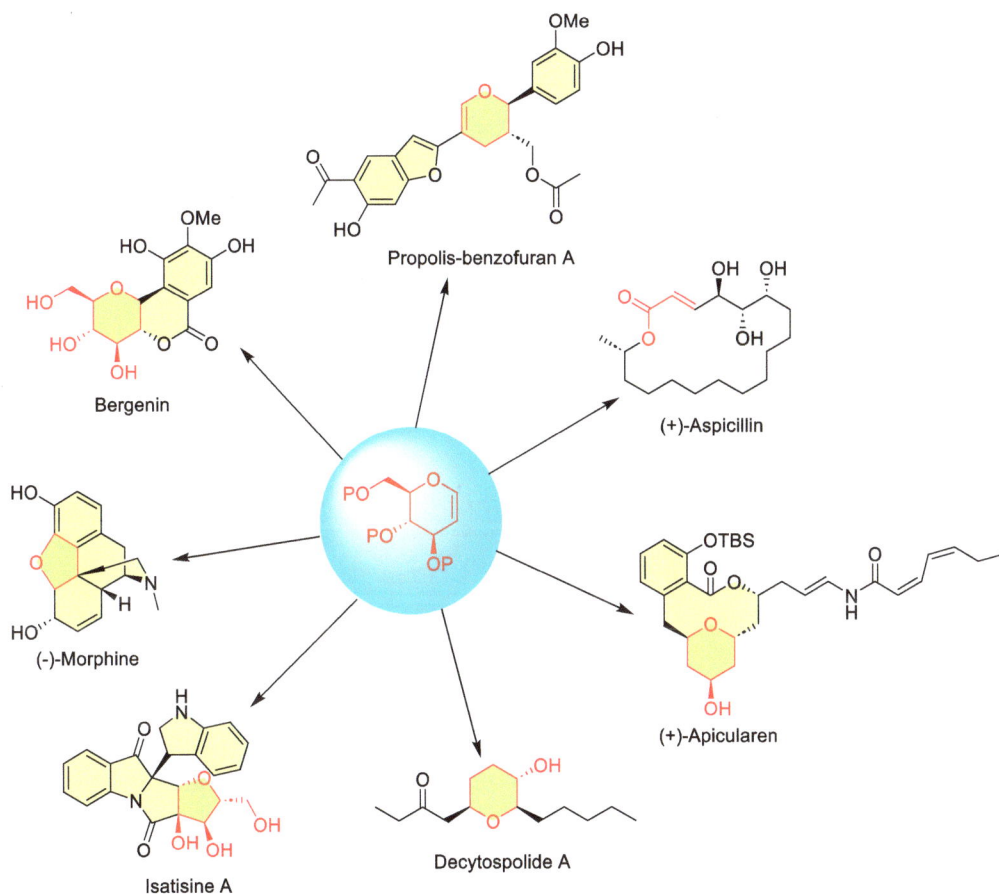

Fig. (2). Glycal-derived natural products.

2-Haloglycals

2-haloglycals have been identified as important building blocks for the further functionalization at the C-2 position of glycals to access C-2-branched sugars and

1,2-annulated sugars employing metal-catalyzed cross-coupling reactions such as Heck, Suzuki, Sonogashira, and carbonylative cross-coupling.

Nowadays, several methods have been reported in the literature for the preparation of 2-haloglycals. For example, Sergio Castillon's group developed a dehydrative reaction condition to synthesize 2-iodoglycals **2**, which applies to various glycals and disaccharides, as depicted in (Scheme **1A**) [8].

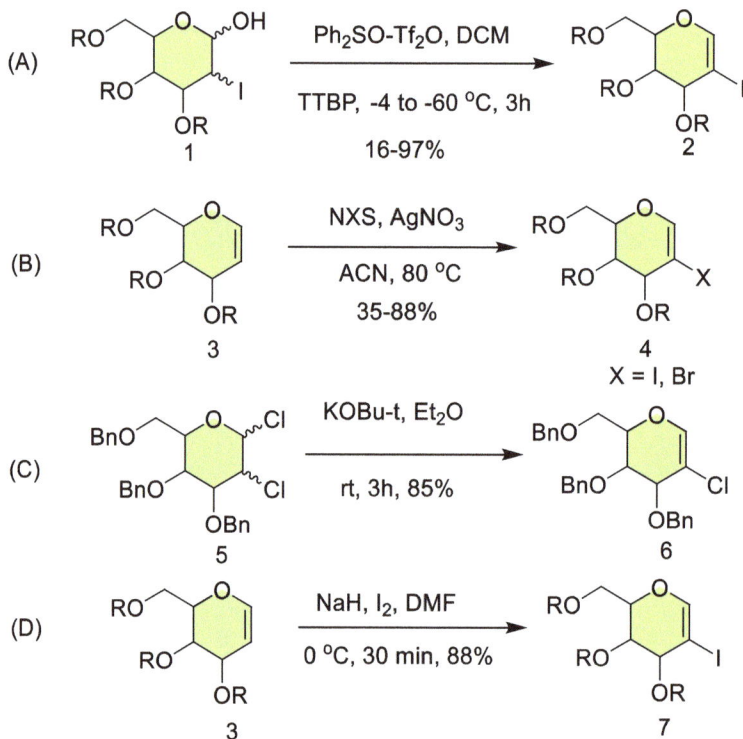

Scheme 1. Different approaches for the preparation of 2-haloglycals.

Later, the research group of Vankar reported a suitable way for the preparation of 2-haloglycals **4** from the glycals **3** using NIS/AgNO$_3$ and NBS/AgNO$_3$ as reagent systems at elevated temperature, as shown in (Scheme **1B**) [9]. This method applies to numerous protecting groups having ester or ether bonds with many glycals like L-rhamnal, D-galactal, D-xylal, and furanoid glycals. The research group of Mukherjee has synthesized 2-chloroglycals **6** by 3,4,6-tri-*O*-benzyl-2-chloro-2-deoxy-glucopyranosyl chloride **5** in the presence of potassium *tert*-butoxide in diethyl ether, as demonstrated in (Scheme **1C**) [10, 11]. Further, this group has synthesized 2-iodo-glycals **7** using molecular iodine and sodium hydride as a reagent system at atmospheric temperature from **3**. Using this method, various 2-iodoglycals having ether, silicon, ester, and acetonide protections were prepared in good to excellent yields, as given in (Scheme **1D**).

The research group of Vankar has synthesized 2-C-branched dienes **10** *via* a cross-coupling reaction of 2-haloglycals **8** in the presence of terminal alkenes **9** with Pd(OAc)$_2$/PPh$_3$ and K$_2$CO$_3$, as depicted in (Scheme **2**) [9]. The authors reported that the activated alkenes, such as acrylates and alkenes, contain electron-withdrawing groups that deliver the corresponding products in excellent yields (up to 94%), but unactivated alkenes like styrene take longer reaction time and lower yields with a mixture of E/Z isomers.

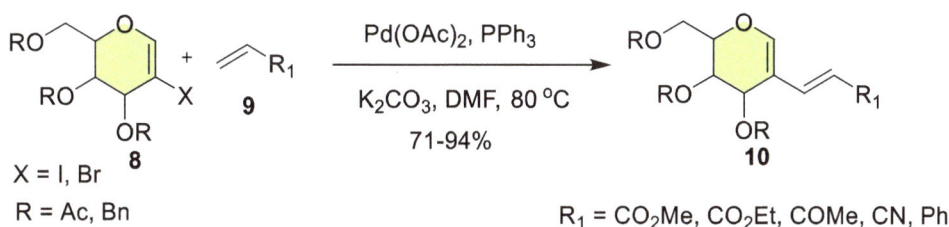

Scheme 2. Heck coupling of 2-haloglycals with terminal alkenes.

Further, the synthetic utility of tri-*O*-acetyl-2-iodoglycals **11** was demonstrated by the successful synthesis of 2,3-unsaturated *O*-glycosides **12** using the Ferrier reaction. The compound **12** on treatment with methyl acrylate using Pd(OAc)$_2$/PPh$_3$ and K$_2$CO$_3$ in DMF provides the C-2-branched sugar **13** with 79% yield (If R = CH$_3$). However, compound **12** underwent an intramolecular Heck reaction (If R = allyl) to give the bicyclic compound **14** in 78% yield, as outlined in (Scheme **3**).

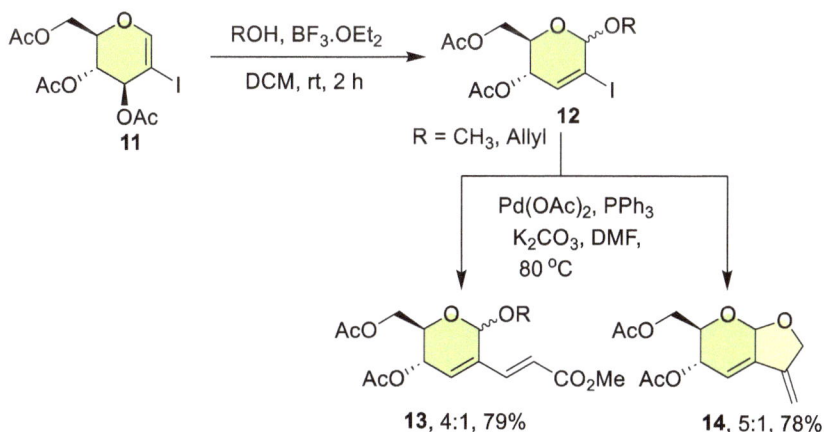

Scheme 3. Application of Ferrier compounds in Heck reaction.

The research group of Boutureira and Davis has utilized 2-iodoglycals **15** with arylboronic acids **16** in a ligand-free Suzuki-Miyaura cross-coupling reaction in aqueous solution for the preparation of 2-C-arylglycals **17**, as depicted in (Scheme **4**) [12]. The researchers disclosed that several substituted arylboronic acids give the desired compounds in good to excellent yields (90-95%). Further, the transformation of C-2-arylglycals **17** into 1,2-disubstituted sugars **18** was carried out *via* epoxidation and ring-opening in an acidic medium with different alcohols. The reduction of the double bond of glycals with Pd/C and H_2 produces 1-deoxy sugar **19** in diastereoisomers.

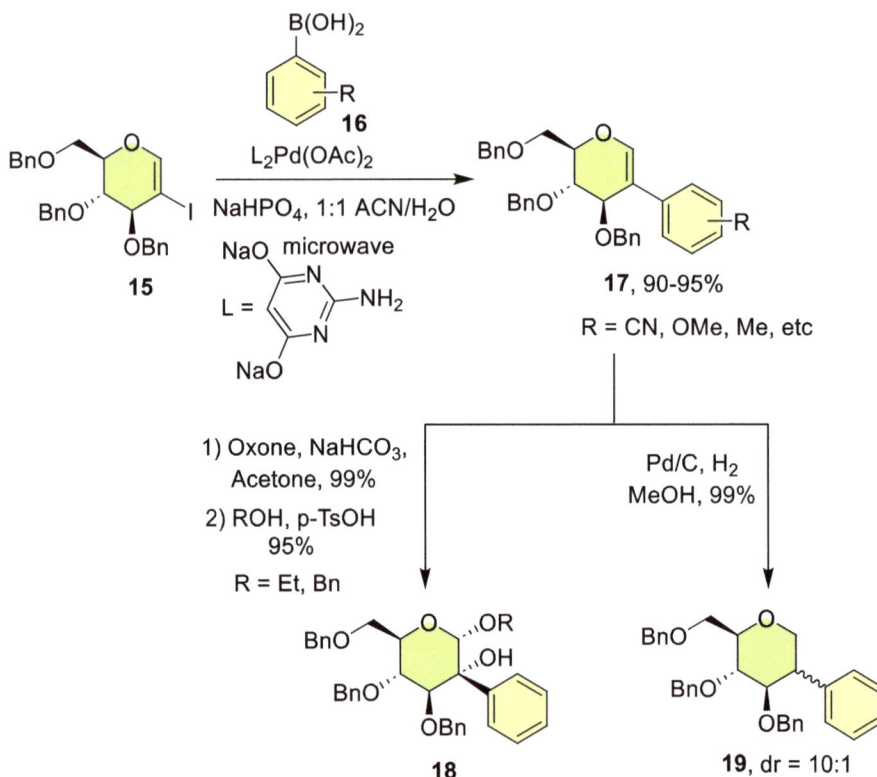

Scheme 4. Synthesis of 2-C-arylglycals and their synthetic transformations.

Samir Messaoudi and his co-workers have explored 2-iodoglycals **20** as potential coupling partners for the formation of thioglycosides **22**, as outlined in (Scheme **5**) [13]. Pd-based reaction in the Buchwald-Hartwig-Migita cross-coupling method was used for the synthesis of both α- and β-thioglycosides. For this transformation, Pd-G3-xantphos was observed as the best catalyst, with base triethylamine and 1,4-dioxane as solvents at 60 °C. The author reported several alkyl thiols **20** treated smoothly with **11** to yield the corresponding 2-C branched sugars **21a-e**. Many glycosyl thiols based on monosaccharides and disaccharides reacted smoothly with 2-iodoglycals to afford various *S*-linked di- and polysaccharides. This process was also applicable to synthesize 2-C-branched glycomimetics when the reaction of 2-iodoglycals was performed with various alkyl thiols under standard conditions.

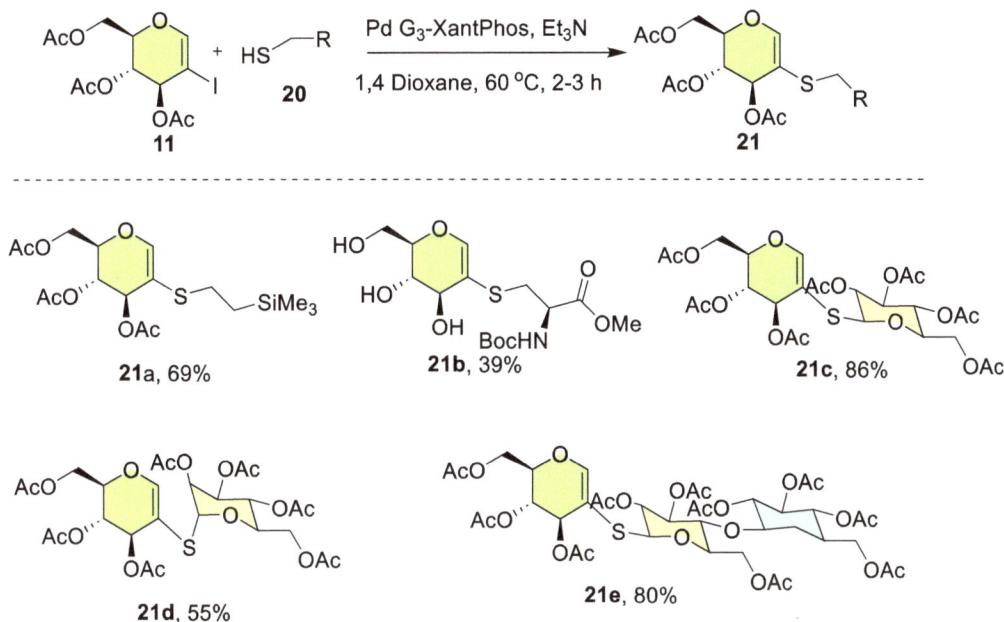

Scheme 5. Synthesis of 2-C-branched thioglycosides.

Angélique Ferry *et al.* have exploited both the protected and unprotected 2-iodoglycals **22** for the synthesis of various 2-cyanoglycals **23**, as demonstrated in (Scheme **6**) [14]. After optimizing the reaction conditions, authors found that the 2-iodoglycal reacted well in the presence of a palladium pre-catalyst, a ligand, a base, and $K_4[Fe(CN)_6]$ in a mixture (1:1) of *t*-BuOH: water at 75 °C under argon atmosphere, which were observed as the best reaction conditions. The protected and unprotected 2-iodoglycals were tolerated under standard conditions to produce the respective 2-cyanoglycals in moderate to good yields.

Further, using this methodology, different 2-C-branched sugars were treated successfully, as illustrated in (Scheme **7**). It was observed that 2-cyanoglycals when treated with Pd/C as a catalyst under H_2 atm, the reduction of both moieties *viz*, cyano group and the double bond of glycal occurred, and the desired product was formed (**24-25**). However, the cyano group was selectively reduced by using $NaBH_4$ of cobalt salt in a catalytic amount to afford the compound **26** in 70% yield. Some other 2-cyanoglycals were also transformed into tetrazole and 2-

amido-based 2-C-branched glycoconjugates (**27-29**). The researchers disclosed that the transformation of 2-iodoglycals **23** into 2-cyanoglycals **23** under Pd-catalyst is a key reaction in the synthesis of diverse 2-C-branched glycoconjugates.

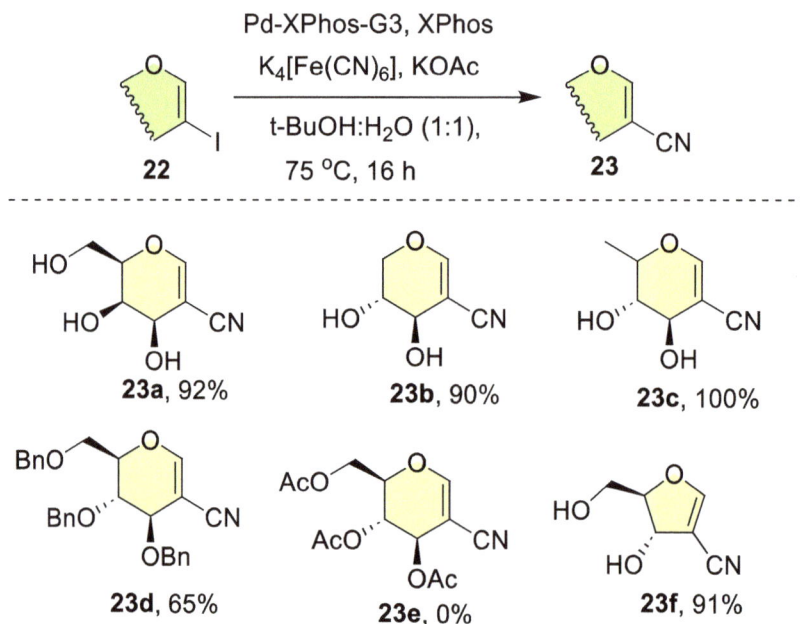

Scheme 6. Synthesis of 2-cyanoglycals from 2-iodoglycals.

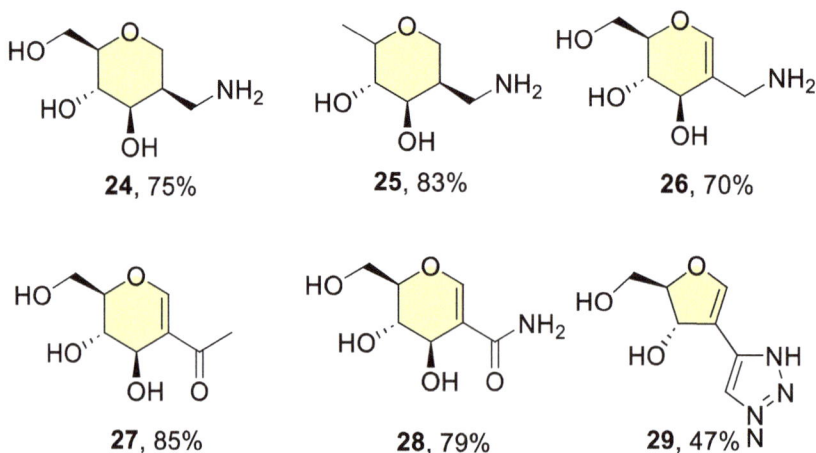

Scheme 7. 2-C-branched glycoconjugates synthesis from 2-cyanoglycals.

In recent years, the introduction of carbon monoxide (CO) into various scaffolds in carbohydrate chemistry has been significantly advanced by the use of transition metal catalysts. Incorporating CO into organic compounds is highly beneficial for chemical industries, particularly in the synthesis of value-added polymers and other consumable products [15]. CO serves as a ligand with a strong affinity for both Pd(0) and Pd(II) due to its ability to function as both a σ-donor and a π-acceptor.

Nadège Lubin-Germain and colleagues utilized 2-iodoglycals for amino carbonylation, employing $Mo(CO)_6$ as the carbonyl source in the presence of a Pd-catalyst to create 2-C-branched glycoconjugates, as shown in (Scheme **8**) [16]. They reported that a stoichiometric amount of metal carbonyl is necessary for this transformation. The optimized carbonylative reaction was compatible with a variety of primary and secondary amines, yielding the corresponding C-2-branched amido-glycals in moderate to good yields.

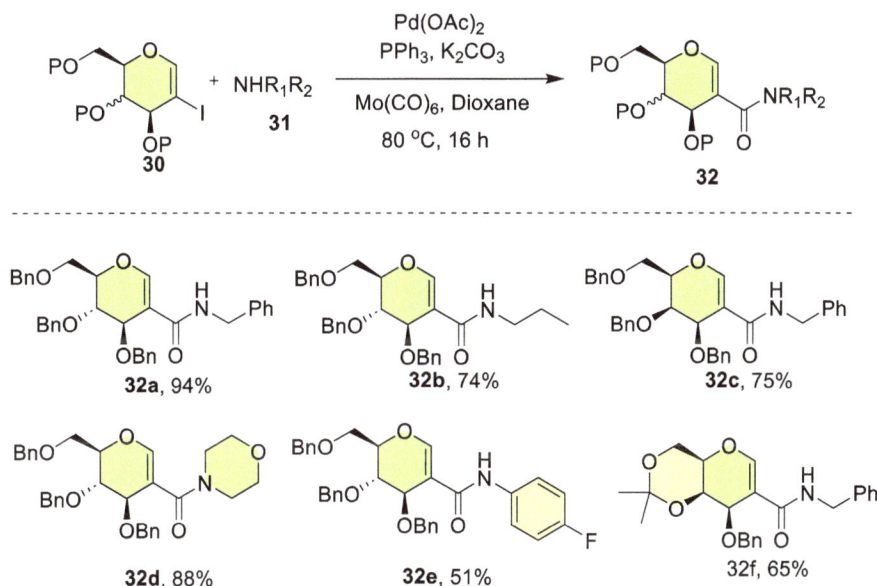

Scheme 8. Carbonylative amination of 2-iodoglycals.

The research group of Hélio A. Stefanihas slightly modified the reaction condition of Nadège Lubin-Germain research group to access amido-glycals **33** and ester-protected glucal **35** under $PdCl_2$ catalysis and $Mo(CO)_6$ used as a carbonyl source [17]. The reaction conditions were found to be suitable for various amines like electronically neutral, electronically rich, and electronically deficient, which, when reacted with 2-iodoglycals **30,** afforded a series of 2-amidoglycals in good yields (Scheme **9A**). Further, it was found that when amines were replaced *via* alcohols **34** under the optimized reaction conditions, the corresponding 2-C branched glucal esters were obtained in good to excellent yields (Scheme **9B**).

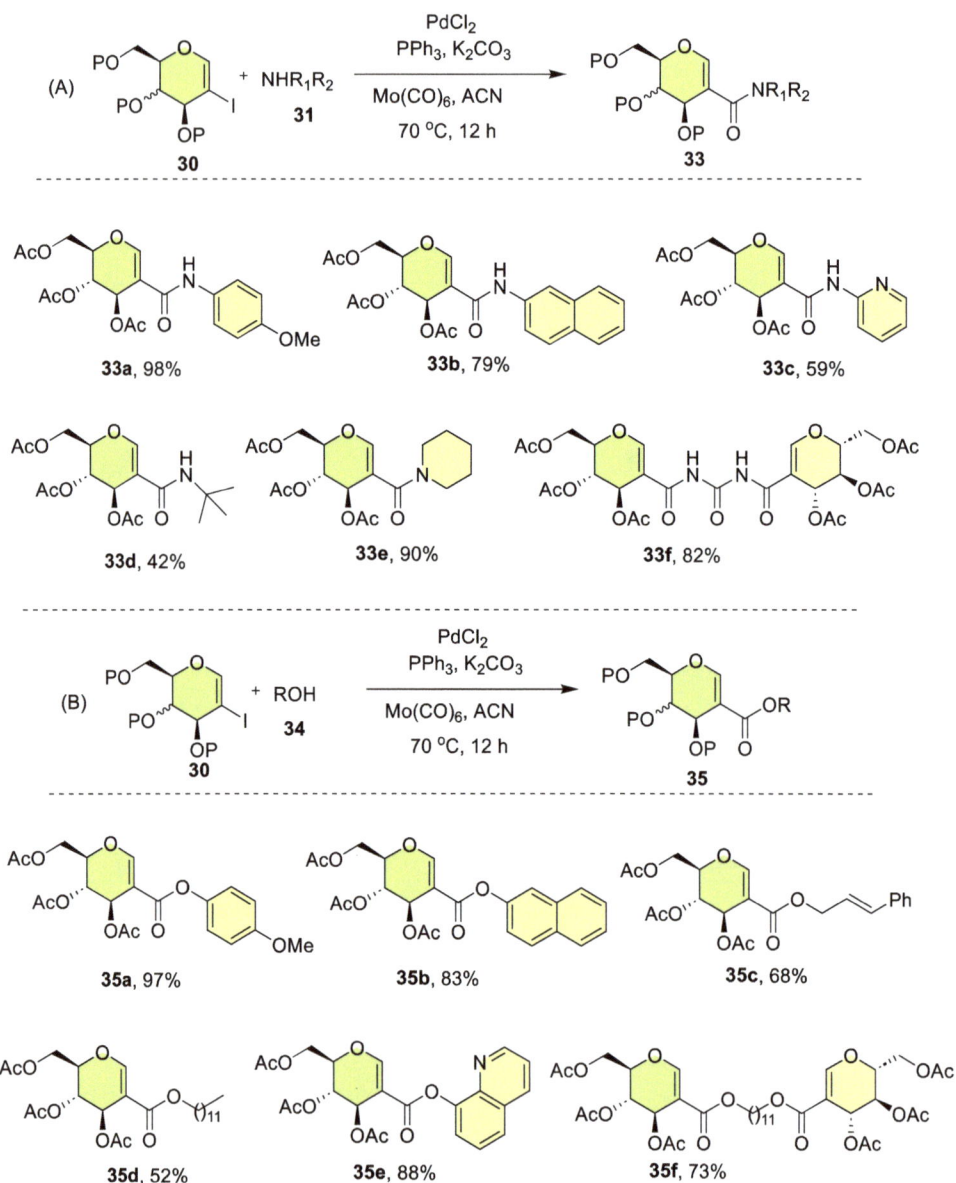

Scheme 9. A) Carbonylative amination of 2-iodoglycal, and **B**) Carbonylative esterification of 2- iodoglycals.

C-H Activation

C-H bond activation reactions play a pivotal role in modern chemical synthesis. It is a direct method for the formation of C-C bonds *via* the activation of the C-H bond. C-C bond formation is the most fundamental linkage in organic chemistry in general and carbohydrate chemistry in particular. In this endeavor, carbohydrate chemists across the globe designed and developed novel strategies for C-H bond activation in glycals, which are described below.

Hussain and his co-workers reported a direct functionalization of glycals *via* Pd-catalyzed cross-coupling with cycloalkenones, as outlined in (Scheme **10**) [18]. The Pd(OAc)$_2$ (10 mol%) was used as a catalyst along with AgOAc (2.5 equiv.) as an oxidant in a mixture of DMF:DMSO (20:1) and found to be the best-optimized reaction conditions. To explore the scope of the reaction, several glycals and cycloalkenones were screened under the optimized conditions, which created functionalized products **38a-d** in good yields.

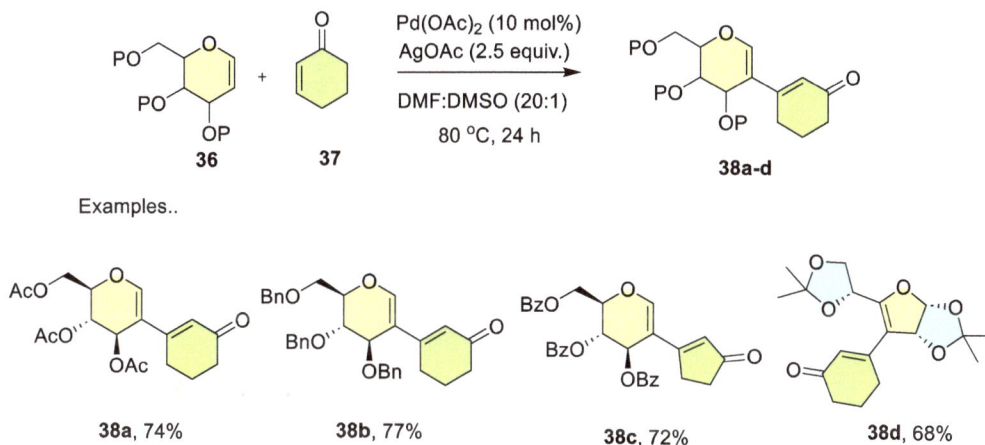

Scheme 10. Direct functionalization of glycals with cycloalkenons.

Phenanthrenones exhibit a wide range of biological activities and are used in various traditional medicines, including expectorants, laxatives, and treatments for skin diseases. The synthetic potential of the products obtained (**38**) when glycals react with cycloalkenones has been demonstrated. This process successfully yields biologically important phenanthrenones (**40**) *via* Diels-Alder cycloaddition, facilitated by the base-promoted ring opening of the sugar moieties.

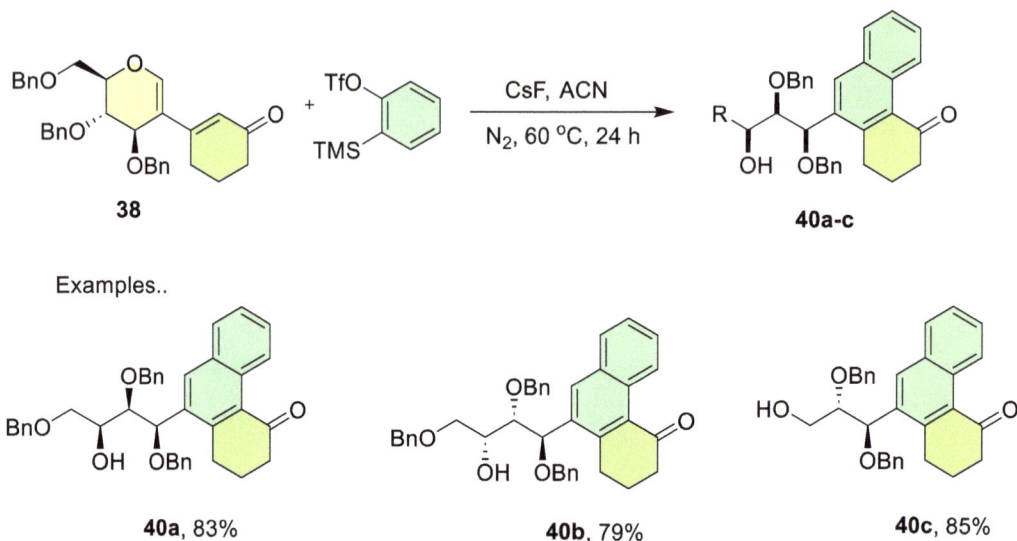

Scheme 11. Synthesis of some biologically important phenanthrenones.

The research group of Samir Messaoudi has reported the direct and diastereoselective method for the synthesis of C2-aryl glycosides, as depicted in (Scheme **12**) [19]. The coupling reaction of acetylated 2,3-pseudoglycals **41** with aryl iodides **42** Pd(OAc)$_2$ (10 mol%) used as a catalyst, AsPh$_3$ (10 mol%) as ligand, and AgTFA as a base in the dioxane solvent at 120 °C afforded diverse C2-aryl glycosides. Numerous aryl iodides bearing diverse substituents reacted smoothly with **41** to form the desired C2-aryl glycosides **43** with high diastereoselectivity in moderate to good yields.

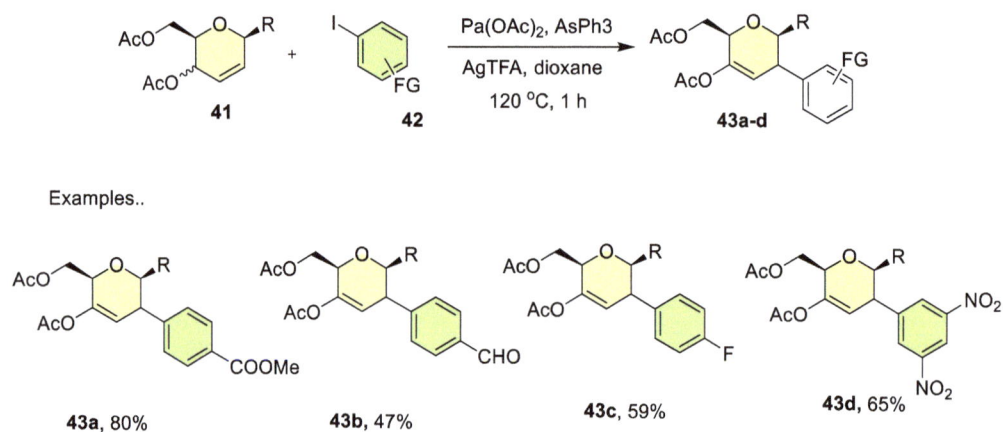

Scheme 12. Pd-catalyzed synthesis of C2-aryl glycosides.

Rajasekaran *et al.* reported the synthesis of 1,2- linearly fused (5,6 and 6,6)-oxa-oxa annulated sugar, as shown in (Scheme **13**) [20]. In this study, 3,4,6-tri-*O*-benzyl D-galactal **15** was first converted into vinyl aldehyde **44** *via* Vilsmeier-Haack reaction. Thereafter, vinylmagnesium bromide was added to vinyl aldehyde at 0 °C, and the resultant hydroxyl group was protected with benzyl bromide in the presence of NaH to yield the compound **45**. After 2 successive steps of the reaction, the compound **46** was obtained, which was then subjected to iodocyclization using *N*-iodosuccinimide (NIS) to form 6-endo iodocyclized product **47**. After two steps of reaction, furan fused sugar molecule **48** was formed with 81% of yield. Using a similar method, two more furan **49** and pyran **50** fused sugar moieties were prepared.

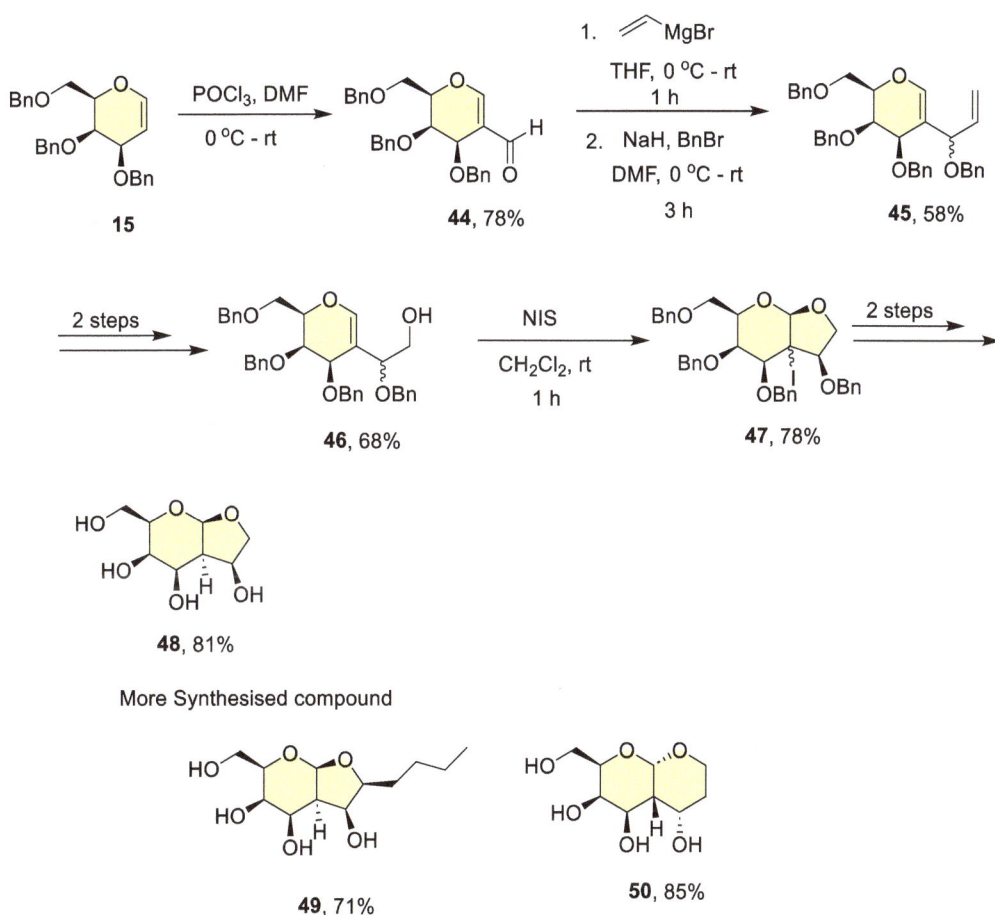

Scheme 13. Synthesis of 1,2- linearly fused (5,6 and 6,6)-oxa-oxa annulated sugar.

Hussain *et al.* have developed a strategy for the synthesis of C-2 branched sugar dienes *via* cross-dehydrogenative coupling of sugar enol ethers **51** with terminal alkenes **52**, as demonstrated in (Scheme **14**) [21]. This method is applicable for both pyran and furan-based enol ethers, which are coupled smoothly with electron-rich and electron-deficient alkene sources, yielding sugar dienes **53** with complete *E*-stereoselectivity. The reaction of glycal with methyl acrylate **52** in the presence of catalyst Pd(OAc)$_2$ (10 mol%) and base AgOAc in solvent DMF/DMSO delivered the desired products **53a-f** in excellent yields. The group has further extended their work using product **53** as a starting material to insert the azide group at the C3 position of glycals [22]. Dienes already installed at the C2 position enhances the reactivity of the C3 position, thereby incorporating the upcoming nucleophilic azide at the C3 position of glycals to yield the desired product in good to excellent amounts **54a-d**, as depicted in (Scheme **15**).

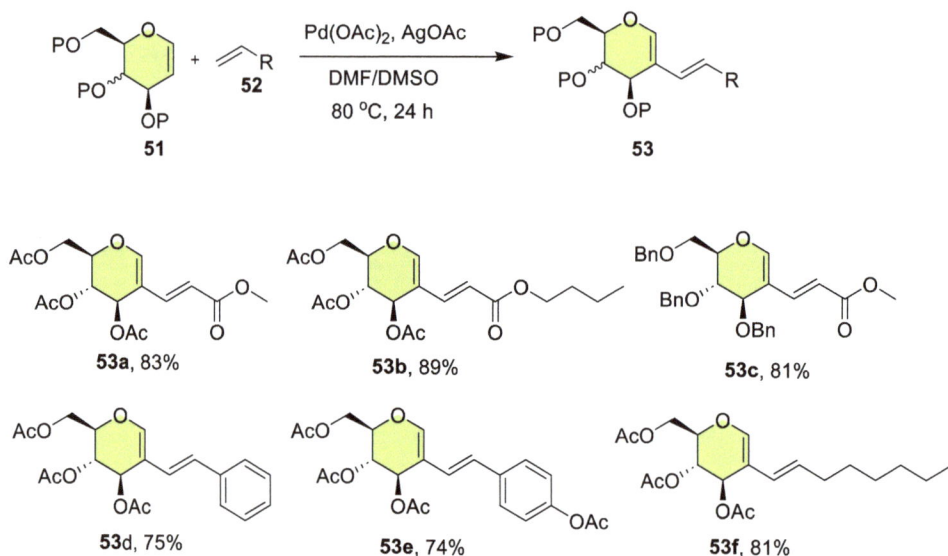

Scheme 14. Synthesis of 2-C-branched sugar dienes.

Another contribution toward the synthesis of 2-C-branched sugars was delivered by Tiwari and co-workers by introducing a thio(aryl)alkyl group at the C-2 position of glycals, as depicted in (Scheme **16A**) [23]. The utilization of pseudoglycals having a double bond at the C2/C3 position of glycals towards the synthesis of C2-functionalized thioglycosides was carried out by conducting the reaction of 2,3-pseudoglycal **55** and thiophenol **56** substrates. The glycal moiety was activated by using BF$_3$·OEt$_2$ (30 mol%) at 30 °C and then reacted with thiophenol to afford the desired C-2 thioglycosidation product **57** in 71–87%

yield. Thiophenol derivatives having halogen atoms on different positions of aromatic rings were examined, and it was found that 2-fluoro-, 4-fluoro-, and 2-bromothiophenol provided the desired products in moderate to good yields. The synthetic utility of the synthesized products was demonstrated in the successful synthesis of various glycal-derived natural products **59a-c**, as shown in (Scheme **16B**).

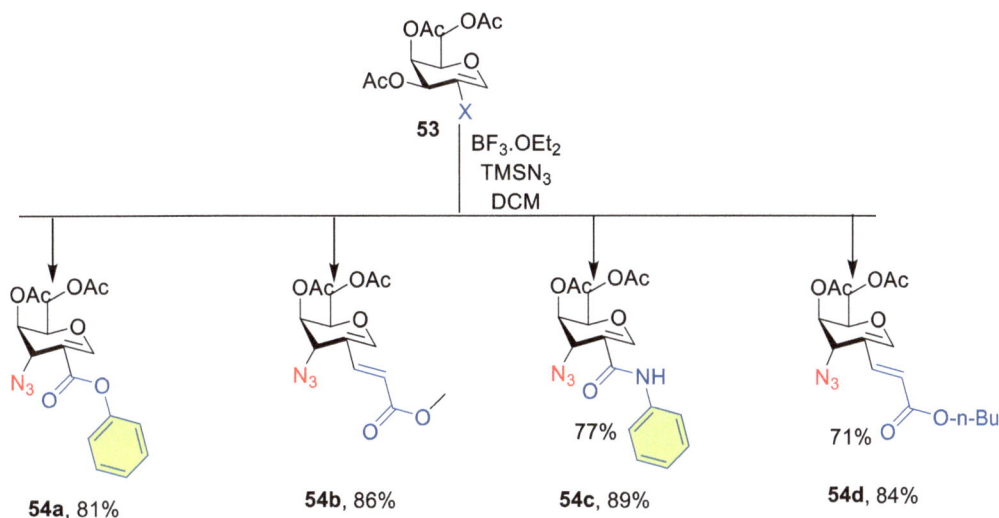

Scheme 15. Incorporation of azide at C3 position.

Besides C-C coupling, the C-O coupling of glycals has also been explored well at the C2 position to access various value-added products. In this regard, Reddy and co-workers reported a Pd-catalyzed oxidative cross-coupling of glycals **11** with aryl carboxylic acids in the presence of stoichiometric amounts of DIB to yield either C-2-aryloxylated **60** or acyloxylated glucal **61**, as shown in (Scheme **17**) [24]. Aryloxylation or acyloxylation of glycal depends on the stoichiometry of the DIB used in the reaction. It could be converted into aryloxylated glucals when 1.0 equiv. of DIB was used. However, acyloxylated glucals were obtained when 2.0 equiv. of DIB was used.

In 2015, Ye and co-workers reported a direct C-H trifluoromethylation of hexose-derived glycals by using an Ir catalyst under photo-redox catalysis, as shown in (Scheme **18**) [25]. In this reaction, [*fac*-Ir^{3+}(ppy)$_3$] was used as a photocatalyst, Umemoto's reagent as a source of trifluoromethyl (CF$_3$) radical, and a household blue LED or sunlight as the light source. Glycals having both electron-withdrawing and electron-donating protecting groups were tolerated well under the standard conditions to afford the desired trifluoromethylated glycals **64**. The above-synthesized compounds were further modified efficiently into pharmaceutical-important 2,3-unsaturated *N*-acetylneuraminic acid (Neu2en) derivatives in the standard conditions.

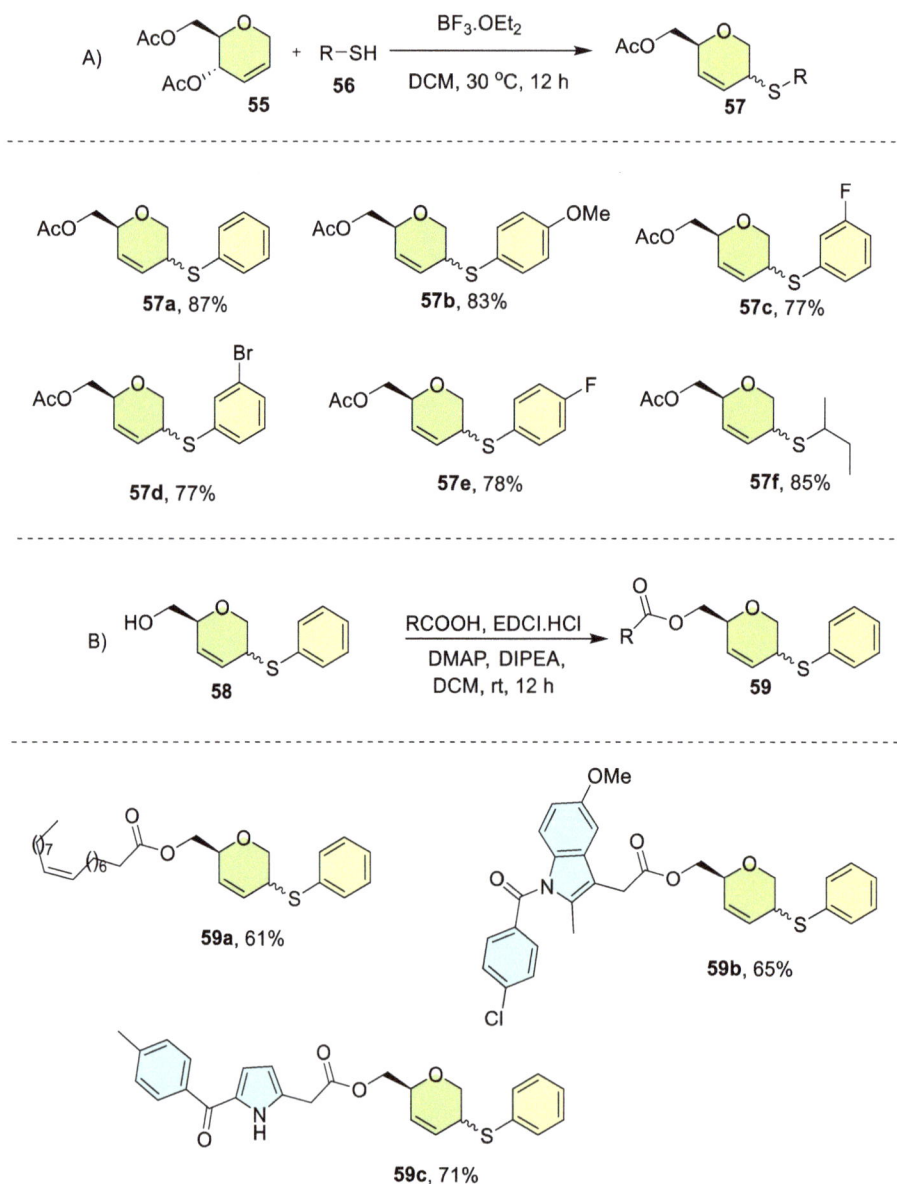

Scheme 16. (**A**) 1-Deoxy-2-thioaryl/alkyl Glycosides (**B**) Late-stage modifications with natural products.

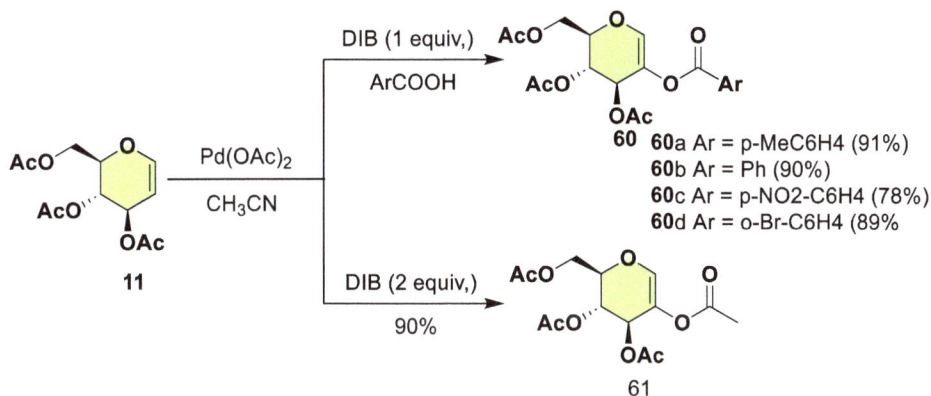

Scheme 17. Representative C-O cross-coupling of glycals at the C-2 position.

A plausible mechanism involving a CF_3 radical and glycosyl oxocarbenium ions is depicted in (Scheme **18**). The reaction was assumed to proceed *via* the formation of an excited state $[fac-Ir^{3+}(ppy)_3]$ with the absorption of light. This high-energy intermediate transfers a single electron to Umemoto's reagent to generate CF_3 radical, which reacts with glycal and generates a glycosyl radical **A**. The glycosyl radical **A** was then oxidized by $[fac-Ir^{4+}(ppy)_3]$ to produce glycosyl oxocarbenium ion **B**, which is equilibrated with the intermediate **C**. Finally, the intermediates **B** or **C** got deprotonated to afford the desired product **64**.

1,2-Annulated Sugars

In recent years, 1,2-annulated sugars have received much attention in carbohydrate chemistry due to their potent biological activities, including glycosidase inhibitors. They are particularly attractive because they offer the possibility to manipulate the anomeric center. In this context, Umbreen and coworkers [26] have disclosed a [2+2] cycloaddition reaction of dichloroketene **67** with acetyl-protected glycals **11** leading to 1,2-annulated sugars **69**, which upon Beckmann rearrangement afforded conformationally fixed glycosamine analogs of pentoses, hexoses, and disaccharides as single regio- and stereoisomers (Scheme **19**).

An interesting example of annulation to the anomeric center has been reported by Jana and Rainier [27]. They found that 2-iodoglycals can easily couple with anilines *via* Suzuki–Miyaura coupling followed by cyclization under two different sets of conditions. Epoxidation and subsequent S_N2-type attack of nitrogen at the anomeric position selectively gives the *cis*-fused indoline **71**; however, NBS exclusively gave the *trans*-fused 1,2-annulated sugar **72**, as depicted in (Scheme **20**).

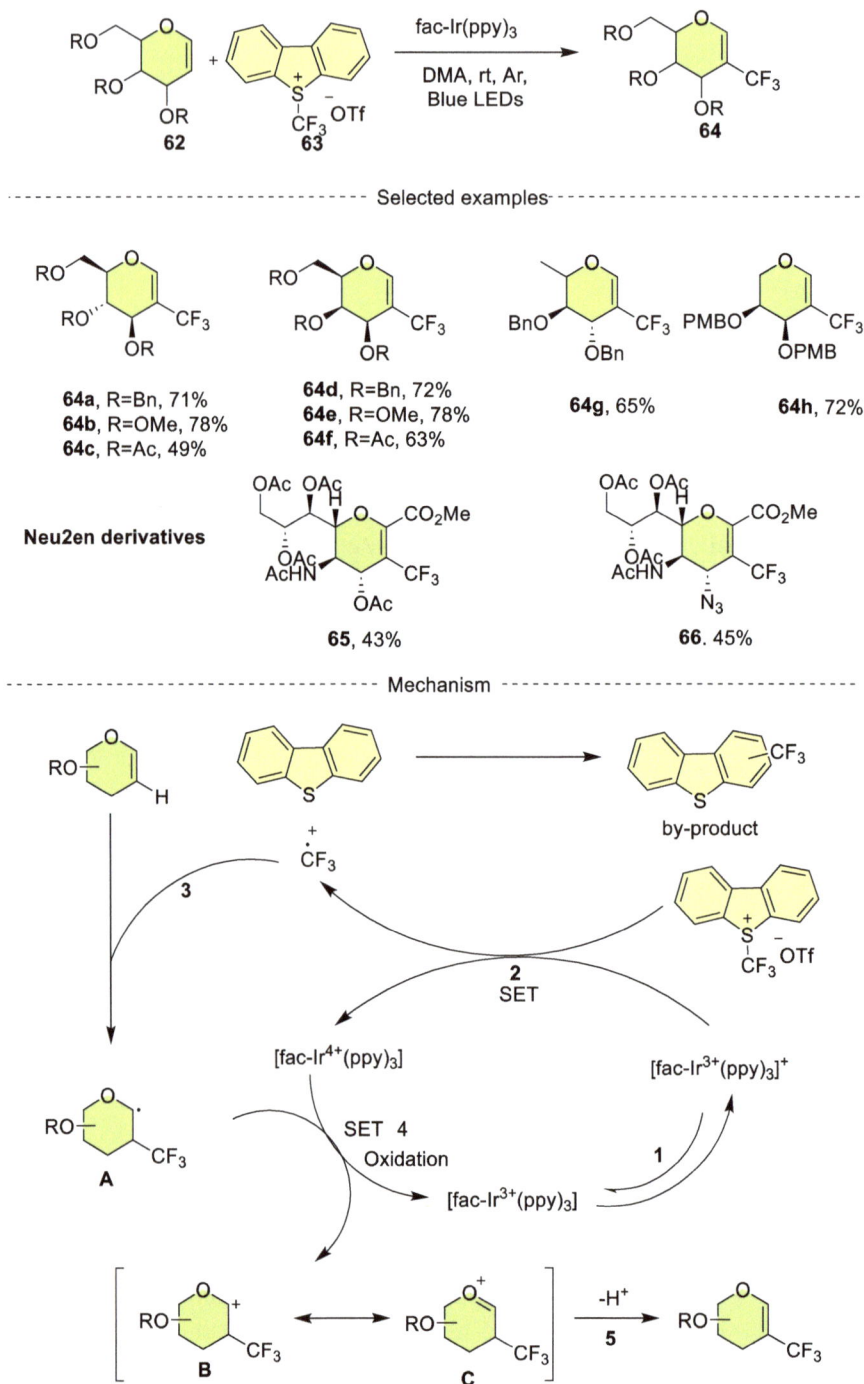

Scheme 18. Photo-redox-catalyzed direct C-H trifluoromethylation of the glycals.

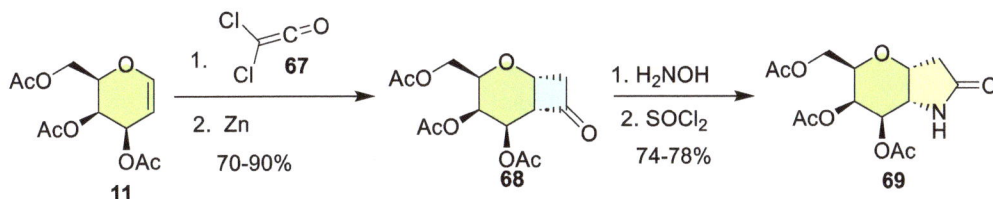

Scheme 19. [2+2] Cycloaddition of ketene with glycals.

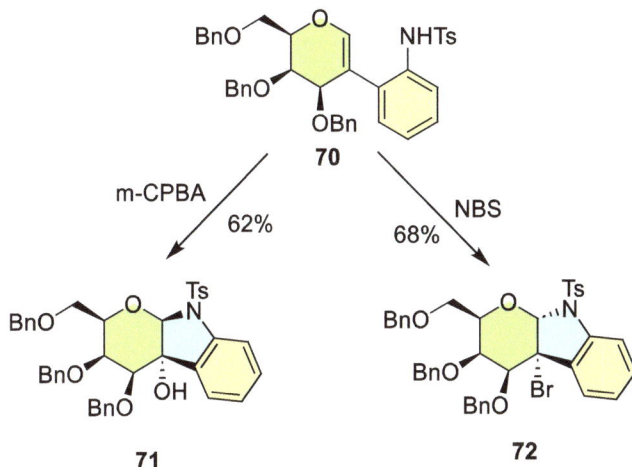

Scheme 20. Glycal-derived 1,2-annulated *cis-* and *trans-*fused indolines.

The studies were directed towards the further exploration of 1,2-annulated sugars, and the research group of D. Mukherjee has reported a one-pot protocol for the construction of glycal-based furo [2,3-b]pyranones **74** through Mn(OAc)$_3$-catalyzed single electron transfer (SET) reactions [28]. The reaction proceeded with glycals using Mn(III)-acetate as a catalyst in the presence of Ac$_2$O and KOAc (radical initiator or stabilizer) at 60 °C under ultrasonic waves, which led to the formation of sugar-fused furo-pyranones with yields up to 78% yield. This method is highly stereo- and regioselective, having diverse substrate scopes with different protecting groups like -OBn, -OAc, -OBz, and -OTBDMS under optimized reaction conditions (Scheme **21**).

Later on, Vankar and co-workers developed a Diels-Alder cycloaddition reaction for the synthesis of annulated *C*-glycosyl amino acids using formyl-galactal **1** [29]. Here, the Diels–Alder reaction of diene **76** with α-nitromethyl acrylate generated *in situ* proceeds in a highly regio- and stereoselective manner to give the cycloadduct **77** as a single stereoisomer. The hydrogenation of **77** in the presence of Raney-Ni as a catalyst, followed by an acetylation reaction, produced sugar-annulated *C*-glycosyl amino ester **78** (Scheme **22**).

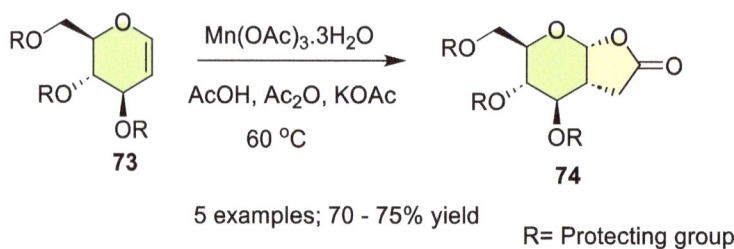

5 examples; 70 - 75% yield

R= Protecting group

Scheme 21. Synthesis of glycals derivedfuro [2,3-*b*] pyran-2-one.

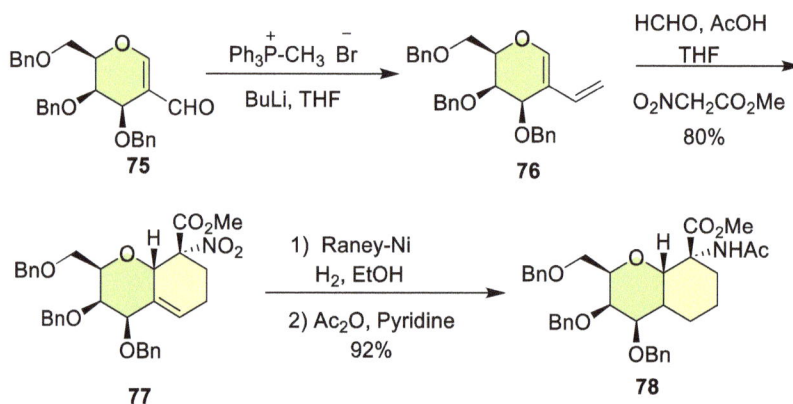

Scheme 22. Synthesis of *C*-glycosyl amino acids.

1-oxadecalins are omnipresent in numerous bioactive natural products that contain a fused 6,6-bicyclic framework. They display various bioactive properties, including antitumor, antifungal, and glycosidase inhibitory activities. This makes the annulated sugars very important potential drug candidates. In this regard, Sun and co-workers [30] synthesized 1-oxadecalins **80a** and **80b** from 2-C-alkenyl glycosides **79** and methyl acrylate *via* Diels–Alder reaction (Scheme **23**). The reaction proceeded in toluene at 110 °C with good yields and high stereoselectivity.

2-Nitroglycals

2-Nitroglycals are useful precursors for the synthesis of various biologically active 2-aminoglycosides, glycoconjugates, and natural products. Several value-added scaffolds such as 2-aminoglycosides, nitro-polyol derivatives, sugar-furan molecules, sugar-pyrrole derivatives, 2,3-aminoglycosides, 2-amino-1-3-dithioglycosides, and many others can be obtained from 2-nitroglycals *via* Michael–type addition, retro-Henry-type reaction, C1/C3-Ferrier rearrangement, and [3+2] *N*-heterocyclic addition. In this regard, Tang *et al.* [31] have disclosed a convenient method for the synthesis of nitro-polyols *via* either a pyridine-promoted scission of the carbon-carbon double bond in 2-nitroglycals or a successive nitration-scission reaction of glycals in one-pot. To demonstrate the synthetic utility of this transformation, the Michael addition reaction of the newly prepared nitro-sugar derivative **83** with methyl acrylate **84** was performed in pyridine water to yield the compound **85**. After 4 steps, compound **85** was converted into bioactive compound 7a-*epi*-(−)-hyacinthacine A1 **86** with 69% yield, as described in (Scheme **24**).

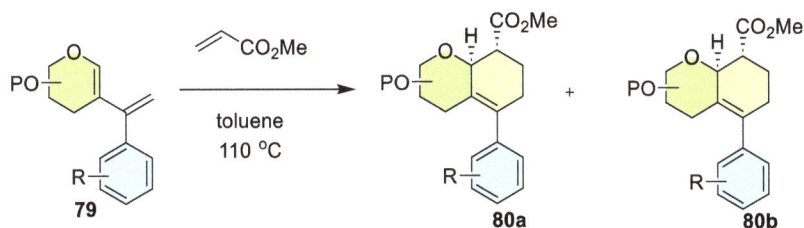

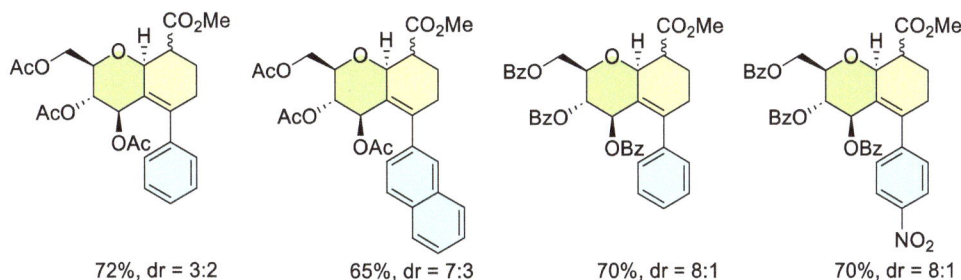

Selected examples..

72%, dr = 3:2   65%, dr = 7:3   70%, dr = 8:1   70%, dr = 8:1

Scheme 23. Synthesis of 1-oxadecalin derivatives **80a** and **80b**.

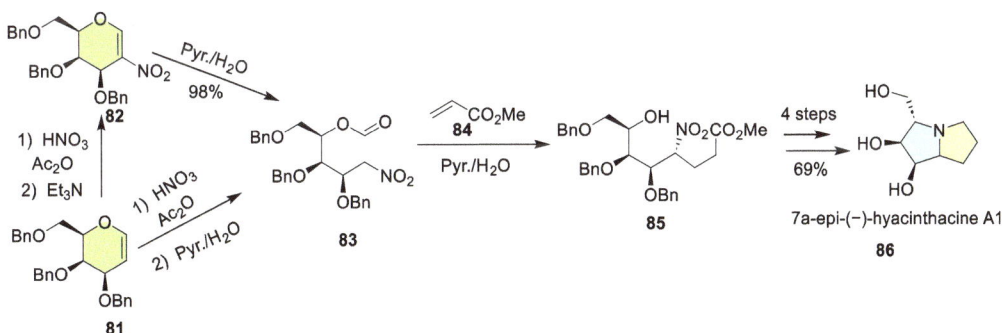

Scheme 24. Synthesis of nitro-polyols and its transformation to 7a-*epi*-(−)-hyacinthacine A1.

Vankar's group [29] made a significant contribution to the chemistry of 2-nitroglycals. They reported a highly stereoselective one-pot method for the synthesis of novel 1,2-annulated sugars comprising of oxa-oxa and oxacarbasugar fused skeletons from 2-nitrogalactal and aryl substituted Baylis-Hillman alcohols (Scheme **25**). The reaction of 2-nitrogalactal **82** with 1.5 equiv. of Baylis-Hillman alcohol **87** using 2.0 equiv. of *t*-BuOK as a base in THF solvent produces the selective reduction of the nitro group and acetylation to afford 2-acetamido-1,2-annulated sugars **88a-f**.

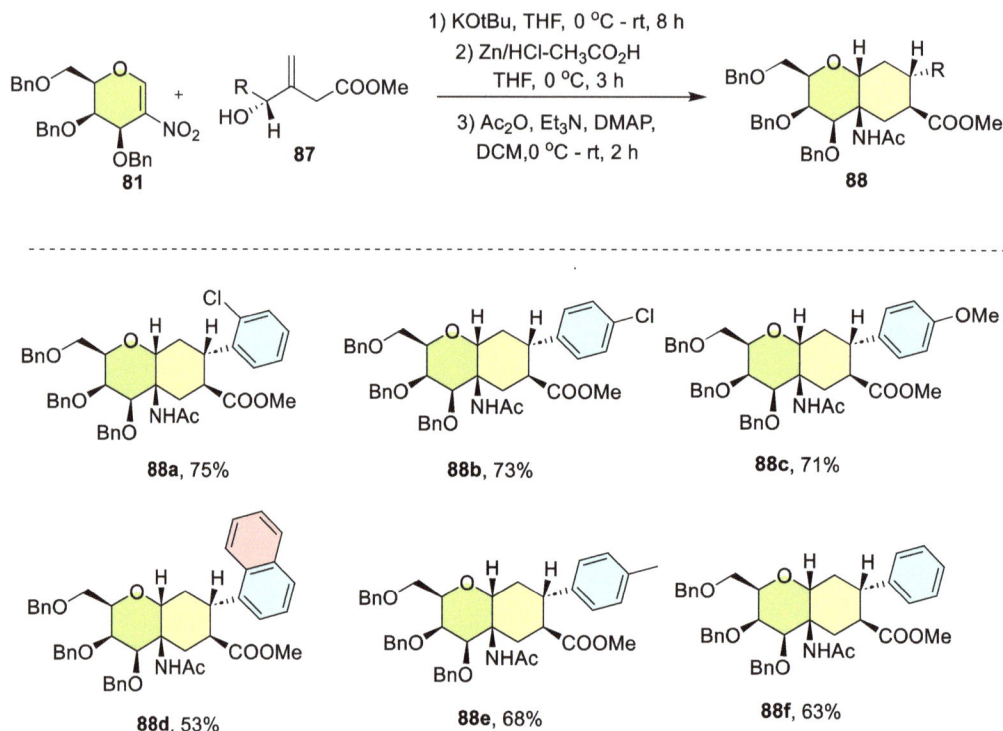

Scheme 25. Synthesis of substituted 1,2-annulated sugars using domino double-Michael addition reaction.

Further studies on 2-nitroglycals were led by Chen and co-workers in 2017 to report the first stereoselective *N*-heterocyclic carbine-catalyzed glycosylation of 2-nitrogalactals **82** with alcohols and phenol to produce various 2-nitroglycosides **90** in good yields, as outlined in (Scheme **26**) [32]. The researchers disclosed that the glycosyl acceptor or other OH group coordinated to a bulky *N*-heterocyclic carbine **89** would attack a 2-nitrogalactal stereoselectively.

Cyclopropanated Sugars

Carbohydrates fused with a cyclopropane ring at their 1,2-positions have been extensively used as a useful precursor in carbohydrate chemistry. They hold a well-defined reactivity towards ring opening to provide either 1,2-annulated sugars or C-2-branched sugars. These substrates can be readily synthesized with various methods such as zinc/copper mediated Simmons-Smith reaction and other strategies [33]. In this regard, Chandrasekaran and coworkers [34] developed a method for highly stereoselective ring opening of a donor-acceptor-based sugar-derived cyclopropane using NIS/NaN$_3$ reagent system followed by coupling with acetylene moieties to form 2-C-branched sugars. The glycal-derived cyclopropane **91** on treatment with NIS/NaN$_3$ provided **92,** which upon further treatment with acetylene using click reaction condition formed disaccharide **94**, as depicted in (Scheme **27**). γ-lactam **95**, an important key unit of many natural products, was successfully synthesized using this methodology, which is an example of 1,2-annulated sugar (Scheme **27**).

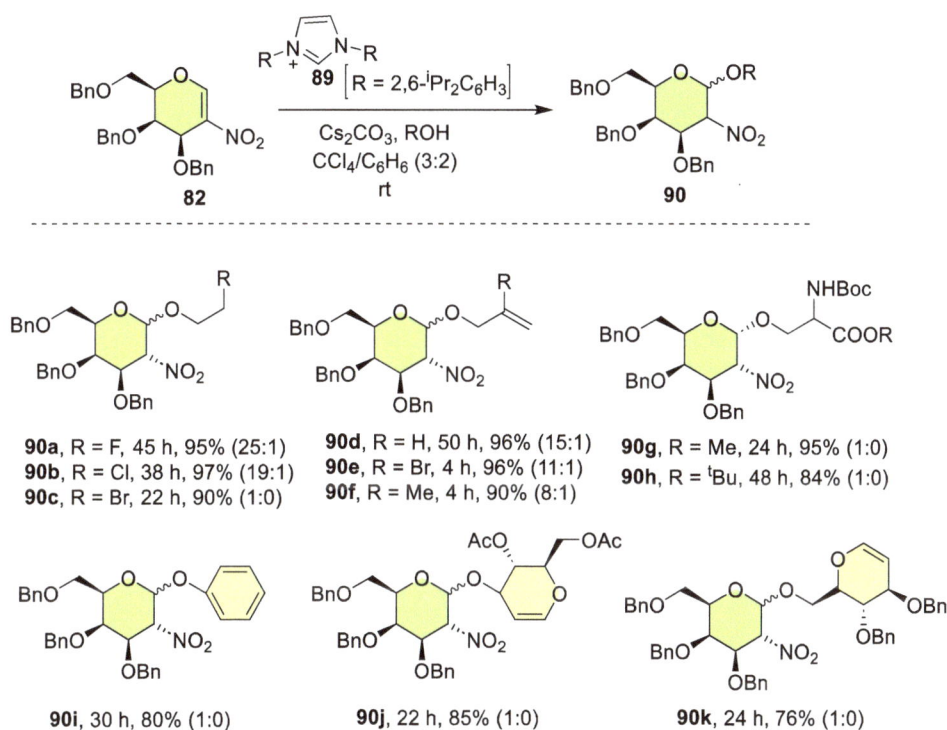

90a, R = F, 45 h, 95% (25:1)
90b, R = Cl, 38 h, 97% (19:1)
90c, R = Br, 22 h, 90% (1:0)

90d, R = H, 50 h, 96% (15:1)
90e, R = Br, 4 h, 96% (11:1)
90f, R = Me, 4 h, 90% (8:1)

90g, R = Me, 24 h, 95% (1:0)
90h, R = ^{t}Bu, 48 h, 84% (1:0)

90i, 30 h, 80% (1:0)

90j, 22 h, 85% (1:0)

90k, 24 h, 76% (1:0)

Scheme 26. NHC-catalyzed stereoselective glycosylation of 2-nitrogalactals.

The halogenated cyclopropane glucals have been well explored by Harvey and their co-workers to synthesize various 2-C-branched and 1,2-annulated sugars using palladium-based cross-coupling reactions [35]. The ring opening of the halogenated cyclopropane ring **95** led to the formation of either oxepine-derived vinyl bromides **96** or six-membered sugars **97** through ring expansion (Scheme **28**). These vinyl bromides were further employed in a Pd-catalyzed coupling reaction for the preparation of compounds **98** and **99**. On the other hand, compound **97** was utilized for intramolecular Heck reaction to afford furo-[2,3-b] oxepine **100**, which is an example of a new class of 1,2-annulated sugars.

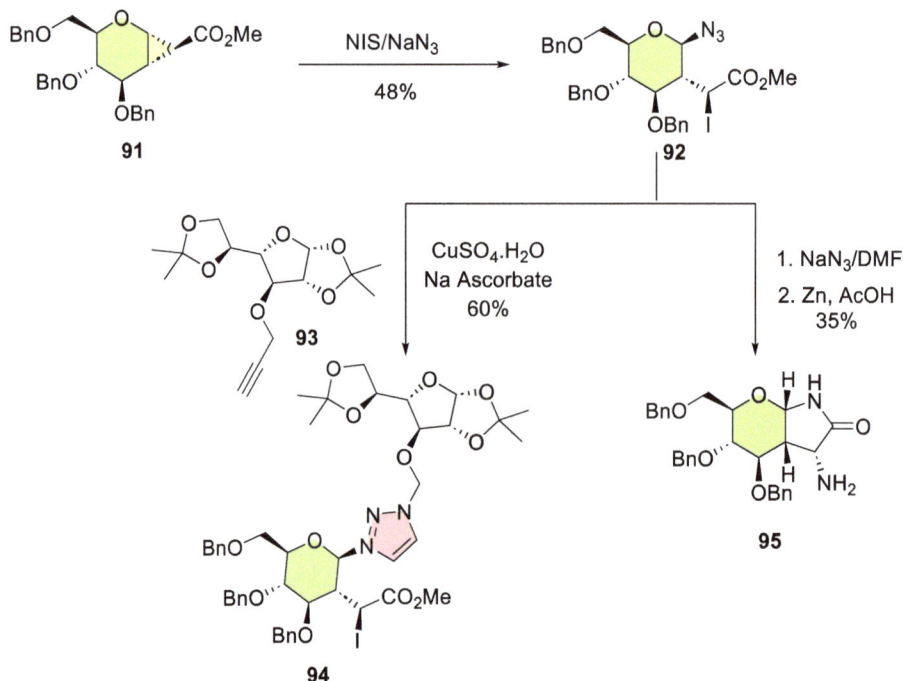

Scheme 27. Conversion of cyclopropane into 2-*C*-branched sugar-derived *N*-glycosides and 1,2-annulated sugars.

Another interesting approach to obtaining 2-*C*-branched glucal derivatives has been developed by Parkan and coworkers [36] using unsubstituted cyclopropane-fused sugars **101**, which undergoes ring opening with *N*-halosuccinimides in the presence of water or methanol to provide corresponding C-2 substituted halo methyl glucal **102**. These compounds **102** are further utilized for the synthesis of interesting *C*-glycosides **104** and **105** in multiple-step reactions, as depicted in (Scheme **29**).

1,2-Cyclopropanated sugars were further exploited by Shao *et al.* for the stereoselective synthesis of 2-*C*-acetonyl-2-deoxy-D-galactosides using 1,2-cyclopropane acetylated sugars **106** as novel glycosyl donors, as depicted in (Scheme **30**) [37]. They disclosed that the BF$_3$·OEt$_2$-catalyzed glycosylation favors *β*-anomeric products while TMSOTf-catalyzed glycosylation prefers *α*-anomers.

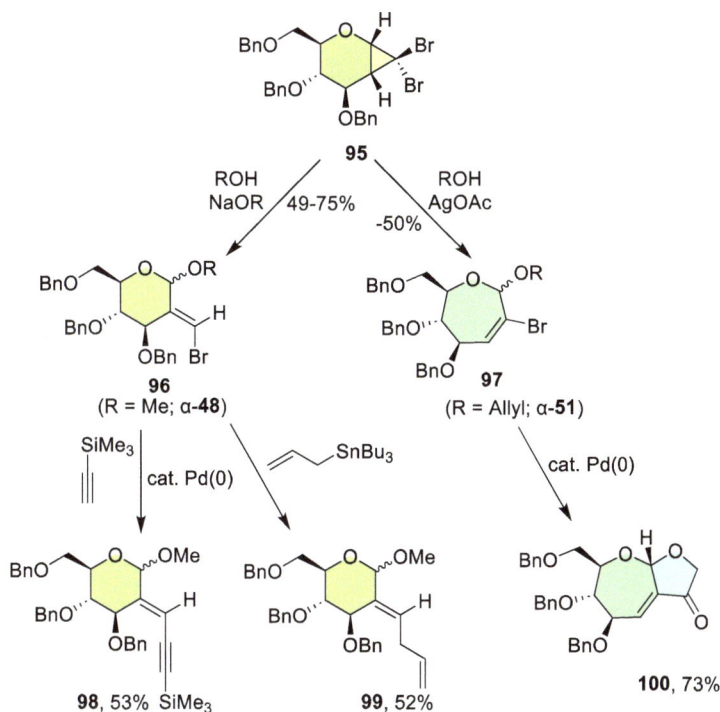

Scheme 28. Conversion of halogenated cyclopropane glucal into 2-*C*-branched sugars and 1,2-annulated sugars.

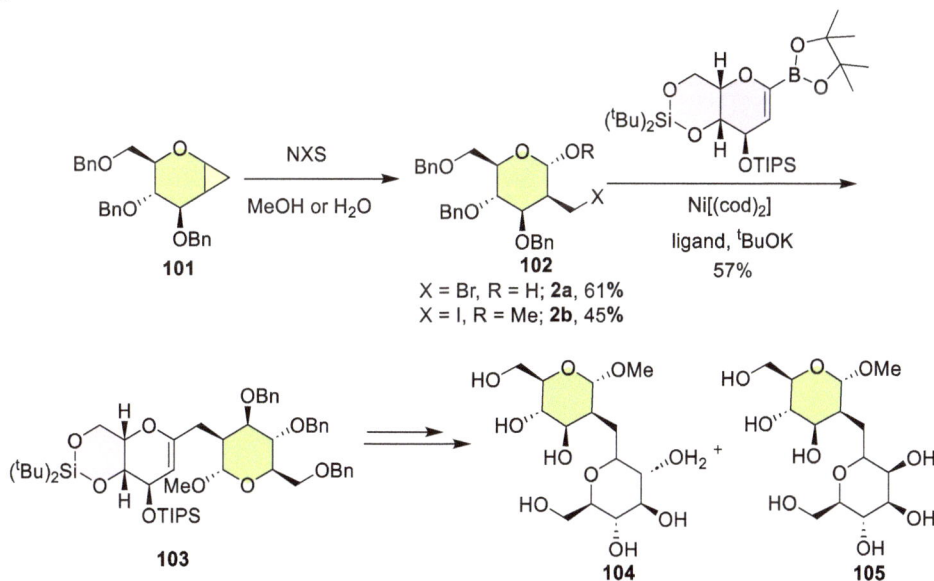

Scheme 29. Conversion of cyclopropanated sugars into fully functionalized 2-branched *C*-glycosides.

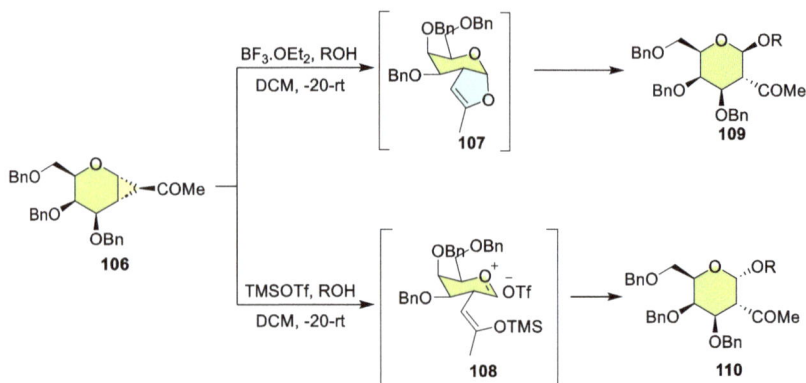

Scheme 30. Stereoselective synthesis of 2-*C*-acetonyl-2-deoxy-D-galactosides from 1,2-cyclopropaneacetylated sugars.

Recently, Harvey *et al.* reported a mechanistic study on the base-promoted ring opening of dibromo cyclopropane glycals **111**, as depicted in (Scheme **31**) [38]. The reaction typically involves the treatment of the cyclopropane ring of **111** with a base in the presence of corresponding alcohol to give *E* as well as *Z*-bromoalkene **113**, as depicted in (Scheme **31A**). A possible mechanistic pathway is described in (Scheme **31B**) for the formation of bromoalkenes stereoselectivity.

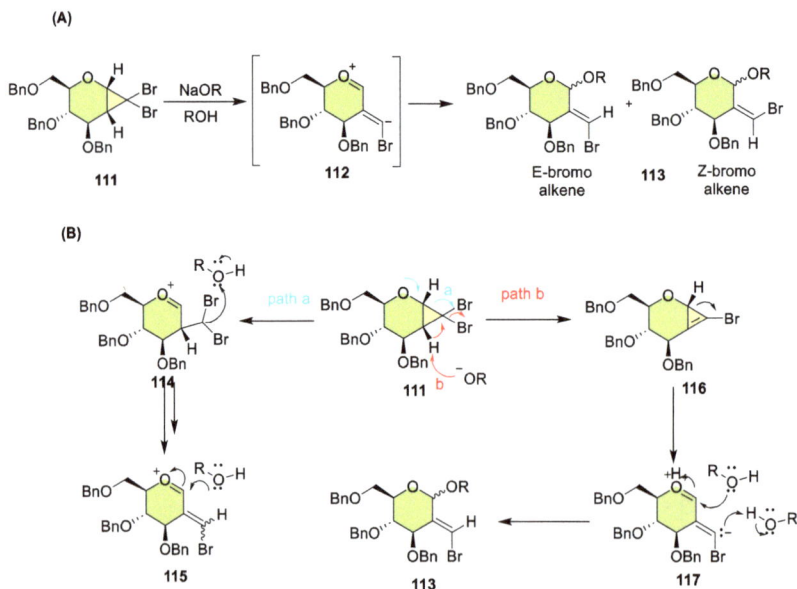

Scheme 31. (A) Ring opening of cyclopropanated glycal and (B) possible mechanistic pathways for the formation of bromoalkenes.

In 2020, Guo and his co-workers reported a highly diastereoselective cyclopropanation reaction of D-glucal and D-galactal derivatives **118** using a simple Rh(II) catalyst in DCM solvent, as shown in (Scheme **32**) [39]. Under the optimized reaction conditions, aryl diazoacetates and fluorinated donor-acceptor diazoalkanes reacted with high efficiencies to give the desired cyclopropane product **120** in a substrate-controlled fashion. Further experiments on these sugar-derived cyclopropanes show a high selectivity for the *β*-anomer **121**.

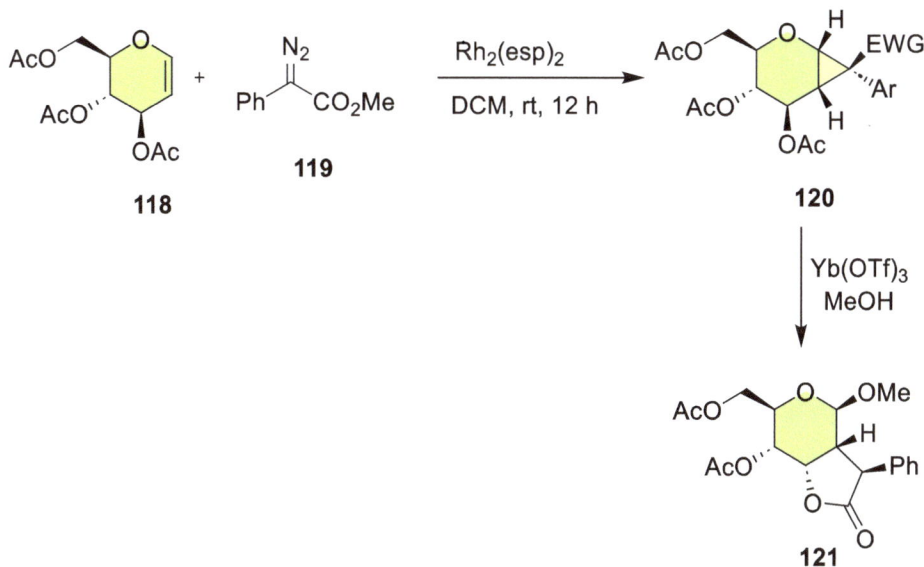

Scheme 32. Cyclopropanation reaction of glycals with aryl diazoacetates.

CONCLUSION

In this chapter, we disclosed different strategies for functionalizing and transforming glycals into C2-branched and annulated sugars. We explored a variety of glycals and their derivatives, including 2-halo glycals, 2-nitro glycals, cyclopropanated glycals, and 2-formyl glycals, each playing a crucial role in synthesizing C-2 branched and annulated sugars. Delving deeper, we delve into the reactivity exhibited at the C-2 position, which serves as the key gateway for the formation of these branched sugars. By understanding the reactivity, we gain insights into the mechanisms underlying the synthesis of complex sugar structures with unique branching patterns. Overall, this chapter provides a comprehensive overview of the strategies and reactions involved in the synthesis of C2 branched and annulated sugars, laying the groundwork for further exploration into the diverse functionalities and applications of these compounds.

ABBREVIATIONS

Pd(OAc)$_2$	Palladium(II) acetate
TTBP 2	4,6-tri-*tert*-butylpyrimidine
PPh$_3$	Triphenylphosphine
Cu (OAc)$_2$	Cupric acetate
K$_2$CO$_3$	Potassium carbonate
DIBAL-H	Diisobutylaluminium hydride
POCl$_3$	Phosphorus oxychloride
***t*-BuLi**	*Tert*-butyllithium
DIPEA	Diisopropyl ethyl amine
LiBr	Lithium bromide
NaHPO$_4$	Sodium phosphate
EDCI.HCl	1-ethyl-3-(3-dimethylaminopropyl)carbodiimide
DCM	Dichloro methane
NIS	*N*-iodosuccinamide
DMAP	4-Dimethylaminopyridine
Mo(CO)$_6$	Molybdenum hexacarbonyl
DIBAL	Diisobutylaluminium hydride
AgTFA	Silver trifluoroacetate
HNO$_3$	Nitric acid
DMDO	Dimethyldioxirne
LDA	Lithium diisopropylamide
TBAF	Tetrabutylammonium hydroxide
MeI	Methyl iodide
NaH	Sodium hydride
PCC	Pyridinium chlorochromate
THF	Tetrahydrofuran
CH$_3$	CN Acetonitrile
KCN	Potassium cyanide
DMSO	Dimethoxy sulfoxide

REFERENCES

[1]   Fischer, E.; Zach, K.; Sitzungsber, K.P.A.W. *Kgl. Preuss. Akad. Wiss.,* **1913**, *27*, 311.

[2]   Ferrier, R.J. *Unsaturated Sugars*; The Carboh, **1980**.

[3]   Ramesh, N.G.; Balasubramanian, K.K. 2-*C*-Formyl Glycals: Emerging Chiral Synthons in Organic Synthesis. *Eur. J. Org. Chem.,* **2003**, *2003*(23), 4477-4487.

[http://dx.doi.org/10.1002/ejoc.200300383]

[4]   Ferrier, R. J.; Hoberg, J. O. *Synthesis and Reactions of Unsaturated Sugars,* **2003**, 55-119.
[http://dx.doi.org/10.1016/S0065-2318(03)58003-9]

[5]   Hussain, N.; Ahmed, A.; Mukherjee, D. 2-Halo Glycals as "Synthon" for 2-*C*-Branched Sugar: Recent
Advances and Applications in Organic Synthesis. *Asian J. Org. Chem.,* **2020**, *9*(6), 882-897.
[http://dx.doi.org/10.1002/ajoc.202000195]

[6]   Vankar, Y.D.; Linker, T. Recent Developments in the Synthesis of 2-*C*-Branched and 1,2-Annulated
Carbohydrates. *Eur. J. Org. Chem.,* **2015**, *2015*(35), 7633-7642.
[http://dx.doi.org/10.1002/ejoc.201501176]

[7]   Somsák, L. Carbanionic reactivity of the anomeric center in carbohydrates. *Chem. Rev.,* **2001**, *101*(1),
81-136.
[http://dx.doi.org/10.1021/cr980007n] [PMID: 11712195]

[8]   Rodríguez, M.A.; Boutureira, O.; Matheu, M.I.; Díaz, Y.; Castillón, S.; Seeberger, P.H. Synthesis of
2-iodoglycals, glycals, and 1,1′-disaccharides from 2-Deoxy-2-iodopyranoses under dehydrative
glycosylation conditions. *J. Org. Chem.,* **2007**, *72*(23), 8998-9001.
[http://dx.doi.org/10.1021/jo701738m] [PMID: 17929873]

[9]   Dharuman, S.; Vankar, Y.D. N-halosuccinimide/AgNO$_3$-efficient reagent systems for one-step
synthesis of 2-haloglycals from glycals: application in the synthesis of 2C-branched sugars *via* Heck
coupling reactions. *Org. Lett.,* **2014**, *16*(4), 1172-1175.
[http://dx.doi.org/10.1021/ol500039s] [PMID: 24499471]

[10]   Sakander, N.; Ahmed, A.; Zargar, I.A.; Mukherjee, D. Base-Mediated Transformation of Glycals to
Their Corresponding Vinyl Iodides and Their Application in the Synthesis of C-3 Enofuranose and
Bicyclic 3,4-Pyran-Fused Furanose. *J. Org. Chem.,* **2023**, *88*(13), 8300-8309.
[http://dx.doi.org/10.1021/acs.joc.3c00302] [PMID: 37315232]

[11]   Hussain, A.; Mukherjee, D. Highly diastereoselective 1,2-dichlorination of glycals using NCS/PPh3:
study of substituent and solvent effects. *Tetrahedron,* **2014**, *70*(6), 1133-1139.
[http://dx.doi.org/10.1016/j.tet.2013.12.088]

[12]   Koester, D.C.; Werz, D.B. Sonogashira–Hagihara reactions of halogenated glycals. *Beilstein J. Org.
Chem.,* **2012**, *8*, 675-682.
[http://dx.doi.org/10.3762/bjoc.8.75] [PMID: 23015813]

[13]   AL-Shuaeeb, R.A.A.; Montoir, D.; Alami, M.; Messaoudi, S. Synthesis of (1 → 2)-S-Linked
Saccharides and S-Linked Glycoconjugates *via* a Palladium-G3-XantPhos Precatalyst Catalysis. *J.
Org. Chem.,* **2017**, *82*(13), 6720-6728.
[http://dx.doi.org/10.1021/acs.joc.7b00861] [PMID: 28598170]

[14]   Malinowski, M.; Van Tran, T.; de Robichon, M.; Lubin-Germain, N.; Ferry, A. Mild Palladium-
Catalyzed Cyanation of Unprotected 2-Iodoglycals in Aqueous Media as Versatile Tool to Access
Diverse C2-Glycoanalogues. *Adv. Synth. Catal.,* **2020**, *362*(5), 1184-1189.
[http://dx.doi.org/10.1002/adsc.201901583]

[15]   Panda, B.; Albano, G. DMF as CO Surrogate in Carbonylation Reactions: Principles and Application
to the Synthesis of Heterocycles. *Catalysts,* **2021**, *11*(12), 1531.
[http://dx.doi.org/10.3390/catal11121531]

[16]   Bordessa, A.; Ferry, A.; Lubin-Germain, N. Access to Complex C2-Branched Glycoconjugates *via*
Palladium-Catalyzed Aminocarbonylation Reaction of 2-Iodoglycals. *J. Org. Chem.,* **2016**, *81*(24),
12459-12465.
[http://dx.doi.org/10.1021/acs.joc.6b02278] [PMID: 27978737]

[17]   Darbem, M.P.; Kanno, K.S.; de Oliveira, I.M.; Esteves, C.H.A.; Pimenta, D.C.; Stefani, H.A.
Synthesis of amidoglucals and glucal esters *via* carbonylative coupling reactions of 2-iodoglucal using
Mo(CO)$_6$ as a CO source. *New J. Chem.,* **2019**, *43*(2), 696-699.

[http://dx.doi.org/10.1039/C8NJ04540B]

[18] Maheshwari, M.; Pandey, R.P.; Hussain, N. Pd-catalyzed direct functionalization of glycals with cycloalkenones: application to the synthesis of chiral phenanthrenones. *Chem. Commun. (Camb.),* **2023**, *59*(5), 627-630.
      [http://dx.doi.org/10.1039/D2CC05255E] [PMID: 36533688]

[19] Ghouilem, J.; Franco, R.; Retailleau, P.; Alami, M.; Gandon, V.; Messaoudi, S. Regio- and diastereoselective Pd-catalyzed synthesis of C2-aryl glycosides. *Chem. Commun. (Camb.),* **2020**, *56*(52), 7175-7178.
      [http://dx.doi.org/10.1039/D0CC02175J] [PMID: 32463030]

[20] Rajasekaran, P.; Ande, C.; Vankar, Y.D. Synthesis of (5,6 & 6,6)-oxa-oxa annulated sugars as glycosidase inhibitors from 2-formyl galactal using iodocyclization as a key step. *ARKIVOC,* **2022**, *2022*(6), 5-23.
      [http://dx.doi.org/10.24820/ark.5550190.p011.809]

[21] Hussain, N.; Babu Tatina, M.; Mukherjee, D. Cross dehydrogenative coupling of sugar enol ethers with terminal alkenes in the synthesis of pseudo-disaccharides, chiral oxadecalins and a conjugated triene. *Org. Biomol. Chem.,* **2018**, *16*(15), 2666-2677.
      [http://dx.doi.org/10.1039/C8OB00168E] [PMID: 29577133]

[22] Pandey, R.P.; Maheshwari, M.; Hussain, N. Synthesis of chiral azides from C-2 substituted glycals and their transformation to C3-glycoconjugates and α-triazolo-naphthalene polyol. *Chem. Commun. (Camb.),* **2023**, *59*(65), 9900-9903.
      [http://dx.doi.org/10.1039/D3CC02423G] [PMID: 37498546]

[23] Hussain, N.; Tiwari, B.; Sharma, M.K.; Pandey, R.P. Site-Selective Lewis Acid Mediated Transformation of Pseudo Glycals to 1-Deoxy-2-thioaryl/alkyl Glycosides. *Synthesis,* **2024**, *56*(6), 944-952.
      [http://dx.doi.org/10.1055/a-2126-0815]

[24] Lahiri, R.; Ansari, A.A.; Vankar, Y.D. Recent developments in design and synthesis of bicyclic azasugars, carbasugars and related molecules as glycosidase inhibitors. *Chem. Soc. Rev.,* **2013**, *42*(12), 5102-5118.
      [http://dx.doi.org/10.1039/c3cs35525j] [PMID: 23535828]

[25] Wang, B.; Xiong, D.C.; Ye, X.S. Direct C–H Trifluoromethylation of Glycals by Photoredox Catalysis. *Org. Lett.,* **2015**, *17*(22), 5698-5701.
      [http://dx.doi.org/10.1021/acs.orglett.5b03016] [PMID: 26562610]

[26] Umbreen, S.; Linker, T. Simple synthesis of conformationally fixed glycosamine analogues by beckmann rearrangement at the carbohydrate ring. *Chemistry,* **2015**, *21*(20), 7340-7344.
      [http://dx.doi.org/10.1002/chem.201406546] [PMID: 25858360]

[27] Jana, S.; Rainier, J.D. The synthesis of indoline and benzofuran scaffolds using a Suzuki-Miyaura coupling/oxidative cyclization strategy. *Org. Lett.,* **2013**, *15*(17), 4426-4429.
      [http://dx.doi.org/10.1021/ol401974v] [PMID: 23952242]

[28] Yousuf, S.K.; Mukherjee, D.; L, M.; Taneja, S.C. Highly regio- and stereoselective one-pot synthesis of carbohydrate-based butyrolactones. *Org. Lett.,* **2011**, *13*(4), 576-579.
      [http://dx.doi.org/10.1021/ol102723c] [PMID: 21244043]

[29] Parasuraman, K.; Chennaiah, A.; Dubbu, S.; Ibrahim Sheriff, A.K.; Vankar, Y.D. Stereoselective synthesis of substituted 1,2-annulated sugars by domino double-Michael addition reaction. *Carbohydr. Res.,* **2019**, *477*, 26-31.
      [http://dx.doi.org/10.1016/j.carres.2019.03.007] [PMID: 30954772]

[30] Liu, J.; Han, P.; Liao, J.X.; Tu, Y.H.; Zhou, H.; Sun, J.S. Palladium-Catalyzed Cross-Coupling of 2-Iodoglycals with *N*-Tosylhydrazones: Access to 2-C-Branched Glycoconjugates and Oxadecalins. *J. Org. Chem.,* **2019**, *84*(14), 9344-9352.
      [http://dx.doi.org/10.1021/acs.joc.9b01056] [PMID: 31264870]

[31] Tang, S.; Xiong, D.C.; Jiang, S.; Ye, X.S. Nitro-polyols *via* Pyridine Promoted C=C Cleavage of 2-Nitroglycals. Application to the Synthesis of (−)-Hyacinthacine A1. *Org. Lett.,* **2016**, *18*(3), 568-571.
[http://dx.doi.org/10.1021/acs.orglett.5b03607] [PMID: 26794249]

[32] Liu, J.L.; Zhang, Y.T.; Liu, H.F.; Zhou, L.; Chen, J. *N*-Heterocyclic Carbene Catalyzed Stereoselective Glycosylation of 2-Nitrogalactals. *Org. Lett.,* **2017**, *19*(19), 5272-5275.
[http://dx.doi.org/10.1021/acs.orglett.7b02543] [PMID: 28906121]

[33] Lebel, H.; Marcoux, J.F.; Molinaro, C.; Charette, A.B. Stereoselective cyclopropanation reactions. *Chem. Rev.,* **2003**, *103*(4), 977-1050.
[http://dx.doi.org/10.1021/cr010007e] [PMID: 12683775]

[34] Haveli, S.D.; Roy, S.; Gautam, V.; Parmar, K.C.; Chandrasekaran, S. Ring opening of activated cyclopropanes with NIS/NaN$_3$: synthesis of C-1 linked pseudodisaccharides. *Tetrahedron,* **2013**, *69*(52), 11138-11143.
[http://dx.doi.org/10.1016/j.tet.2013.11.005]

[35] Moore, P.W.; Schuster, J.K.; Hewitt, R.J.; Stone, M.R.L.; Teesdale-Spittle, P.H.; Harvey, J.E. Divergent synthesis of 2-C-branched pyranosides and oxepines from 1,2-*gem*-dibromocyclopropyl carbohydrates. *Tetrahedron,* **2014**, *70*(39), 7032-7043.
[http://dx.doi.org/10.1016/j.tet.2014.06.069]

[36] Oroszova, B.; Choutka, J.; Pohl, R.; Parkan, K. Modular Stereoselective Synthesis of (1→2)-C-Glycosides based on the sp^2–sp^3 Suzuki–Miyaura Reaction. *Chemistry,* **2015**, *21*, 7043-7047.
[http://dx.doi.org/10.1002/chem.201406591] [PMID: 25801323]

[37] Tian, Q.; Xu, L.; Ma, X.; Zou, W.; Shao, H. Stereoselective synthesis of 2-C-acetonyl-2-deoxy- D-galactosides using 1,2-cyclopropaneacetylated sugar as novel glycosyl donor. *Org. Lett.,* **2010**, *12*(3), 540-543.
[http://dx.doi.org/10.1021/ol902732w] [PMID: 20041707]

[38] Lepage, R.J.; Moore, P.W.; Hewitt, R.J.; Teesdale-Spittle, P.H.; Krenske, E.H.; Harvey, J.E. Mechanistic Studies on the Base-Promoted Ring Opening of Glycal-Derived *gem*-Dibromocyclopropanes. *J. Org. Chem.,* **2022**, *87*(1), 301-315.
[http://dx.doi.org/10.1021/acs.joc.1c02366] [PMID: 34932347]

[39] Guo, Y.; Pei, C.; Koenigs, R.M. Substrate-Controlled Cyclopropanation Reactions of Glycals with Aryl Diazoacetates. *ChemCatChem,* **2020**, *12*(16), 4014-4018.
[http://dx.doi.org/10.1002/cctc.202000569]

CHAPTER 5

Total Synthesis of Natural Products and Medicinally Important Molecules from Glycals

Norein Sakander[1] and **Qazi Naveed Ahmed**[2,*]

[1] *Natural Products and Medicinal Chemistry Division, CSIR-Indian Institute of Integrative Medicine, Canal Road, Jammu-180001, India*

[2] *Academy of Scientific and Innovative Research (AcSIR), Ghaziabad-201002, India*

Abstract: Glycals have been widely used as a versatile building block for the synthesis of C-glycosides and branched sugars and the total synthesis of natural products and biologically active molecules. The versatility of glycals is due to their easy availability and the presence of a ring oxygen connected to a double bond. The inherent chirality of glycals also makes them valuable for the synthesis of various natural products and pharmaceuticals. This chapter provides a detailed overview of the progress made in synthesizing natural products and important molecules derived from glycals.

Keywords: Bradyrhizose, Carbohydrates, Catalysis, Cycloaddition, Drugs, Diospongin, Epoxide, Glycals, Glycosides, Glucopyranoside, Hemicetal, Natural products, Reblastatin, Synthon, Stereoselectivity, Total synthesis, Tricyclic flavonoid, Takai olefination, Yamaguchi esterification, α,β-unsaturated ketones.

INTRODUCTION

The field of organic chemistry encompasses a complex and highly rewarding area known as the total synthesis of natural products and medicinally vital scaffolds derived from glycals [1]. Sugar enol ethers, which serve as versatile building blocks, provide an exceptional foundation for the creation of intricate structures present in natural compounds and substances of pharmaceutical significance. This synthetic approach entails utilizing the distinctive reactivity and structural characteristics of glycals, which are derivatives of carbohydrates that possess a double bond between two carbon atoms in a ring structure [2]. By capitalizing on their inherent flexibility, tolerance towards functional groups, and ability to control stereochemistry, chemists undertake complex routes to create elaborate molecular structures. The importance of total synthesis resides in its ability to co-

* **Corresponding author Qazi Naveed Ahmed:** Academy of Scientific and Innovative Research (AcSIR), Ghaziabad-201002, India; E-mail: naqazi@iiim.res.in

mpletely replicate, within a controlled laboratory environment, natural molecules that possess a diverse array of biological effects and therapeutic possibilities. Through the utilization of glycals as initial substances, scientists navigate through strategic chemical conversions, such as glycosylation, stereoselective reactions, and manipulation of functional groups, to construct intricate frameworks of natural products [3]. The synthesis of these compounds not only allows for a more profound comprehension of their structures and biological functions but also provides opportunities for the creation of innovative medicinal agents. By synthesizing natural products and medically important molecules from glycals, researchers strive to uncover fresh avenues for drug discovery and therapeutic advancements, potentially addressing unmet medical requirements and improving human well-being [1]. The dynamic interplay between organic synthesis and medicinal chemistry is exemplified by the synergy between the intricate chemistry of glycals and the pursuit of unlocking the therapeutic potential of natural compounds. This synergy has led to groundbreaking discoveries in the realm of drug development and molecular design.

Additionally, glycals have been widely employed as the starting material for the creation of 1,2-annulated sugars [4] and additionally for the synthesis of a broad spectrum of natural products, as depicted in Fig. (**1**) [5]. Glycals and their derivatives are used not only to synthesize natural products but also to produce SGLT-2 inhibitors like dapagliflozin, canagliflozin, and empagliflozin [6]. Glycals are now more useful as a building block in organic synthesis due to the production of these naturally occurring or artificially created therapeutically relevant compounds. In the literature, a few review articles showed the importance of glycals in the transformation to natural products and biologically relevant molecules [3a, 3h]. Here, we provide a summary of the use of glycals in the overall synthesis of natural products and scaffolds with medical significance.

Synthesis of Diospongin A

The bioactive steroid Diospongin A was obtained from marine sponges that are members of the Diospongia genus. It is a simple 2,4,6-tri substituted pyran natural product isolated from the rhizomes of *Dioscorea spongiosa* exhibiting potent anti-osteoporotic activity [7]. A significant advancement in organic chemistry has been made with the effective synthesis of diospongin A. This is a significant accomplishment because this natural substance has a complex and fascinating structure. Diospongin A was synthesized entirely from tri-*O*-acetyl-D-glucal through a complex series of processes. Tri-*O*-acetyl-D-glucal **1** is first transformed into α,β-unsaturated ketone **2** over three steps, which is then subjected to the Michael addition of phenyl lithium that is catalyzed by Cu to yield diastereomeric ketones **3** and **3a** in a 1:1.2 ratio. The stereoselective keto reduction of compound **3** is achieved using L-selectride followed by MOM-protection using MOM chloride in the presence of DIPEA in DCM, giving **4**. *Ent*-diospongin A **7** is produced in good yields by dehydroxylation using Im$_2$CS, AIBN, and Bu$_3$SnH in toluene, followed by side chain expansion over 4 steps, as illustrated in (Scheme **1**) [8].

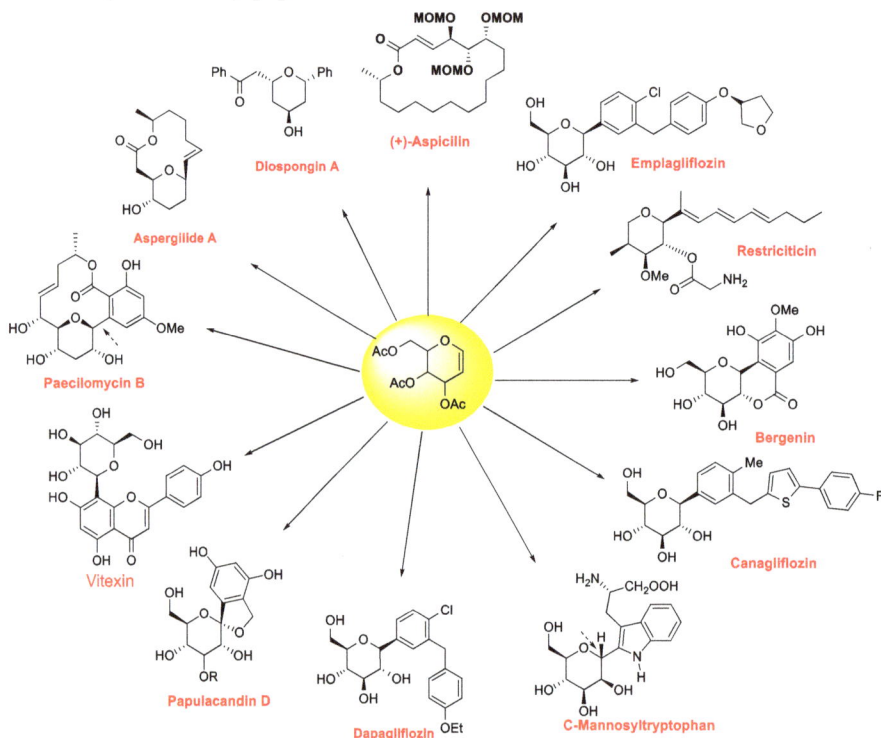

Fig. (1). A few representative examples of natural products derived from glycals.

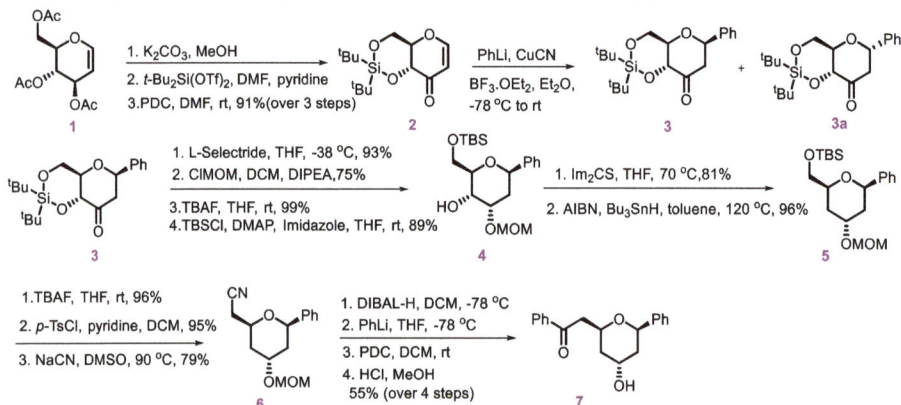

Scheme 1. Synthesis of Diospongin A.

Synthesis of Bradyrhizose

Bradyrhizose **39**, a bicyclic monosaccharide, is a crucial part of the *O*-antigen side chain of lipopolysaccharides found in *bradyrhizobium* species. Through a symbiotic interaction, it helps tropical leguminous plants fix nitrogen [9]. Bradyrhizose is chemically composed of a trans-fused pyranose ring and an inositol-type backbone. Yu and colleagues reported the first synthesis of bradyrhizose from tri-*O*-acetyl-D-glucal, as depicted in (Scheme **2**) [10]. Glycal **1** is reacted with methyl acrylate in the presence of Cu(OAc)$_2$ under Pd catalysis to get conjugated glycal **9**. Next, in three steps, α,β-unsaturated ketone **10** is created by regio- and stereoselectively converting glycal **9**. Compound **11** was produced by reduction, selective TBDPS, and acetylation protection. Selectively iodinating the primary OH group after acetonide deprotection of **11** resulted in compound **12**. The process of selectively oxidizing equatorial 4-OH and then reacting with MeLi yielded diastereoselective exoglycal **13**. After using a protection-deprotection approach to convert **13** to **14**, cyclohexanone **15** was produced through a reaction with sodium chloride and mercuric acetate. The reduction of the keto group in **15,** followed by benzoyl protection, gave **16**. Next, Dess-Martin periodinane oxidation, followed by NaBH$_4$ reduction, epimerized at 5-OH and lastly transformed into **17** by TBDMS protection. The compound **17** was changed into triol **18** by acetyl deprotection, followed by stereoselective dihydroxylation. Pyranose **19** was produced by deprotecting **18** to prevent epimerization, oxidizing it with TEMPO/TCCA, and then reducing it with DIBAL-H. The necessary bradyrhizose was generated in good yield by debenzylating **19** using Pd/C.

Synthesis of Bergenin

The Saxifragaceae family of plants, which includes the significant medicinal herb *Bergenia ligulata*, contains the active secondary metabolite bergenin, which is used in conventional medicine [11]. Bergenin is also a good free radical scavenger and presents excellent antioxidant activity [12, 13]. The pioneer preparation of Bergenin **29** from D-glucal **1a** has been demonstrated by Parkan and his co-workers, as illustrated in (Scheme **3**) [14]. The protection of third position hydroxy group with TIPS and 4,6-hydroxyl groups of D-glucal with di-*tert*-butyl silyl triflate is the first step. D-glucal pinacol boronate **23** is the result of hydrolyzing the C-B bond after treating the protected D-glucal **21** with *t*-BuLi in the presence of borane. Aryl glycoside **25** was obtained by reacting D-glucal pinacol boronate **23** with aglycon **24**, which was catalyzed by palladium. Aryl glycoside **25** was hydroborated with BH$_3$•Me$_2$S and then oxidized to produce glycoside **26**. Further oxidation of the glycoside's benzylic alcohol moiety in **26** and then cyclization produced lactol **27,** which underwent additional oxidation to provide lactone **28**. Ultimately, Bergenin **29** was produced by disilylation and benzyl group deprotection of **28**.

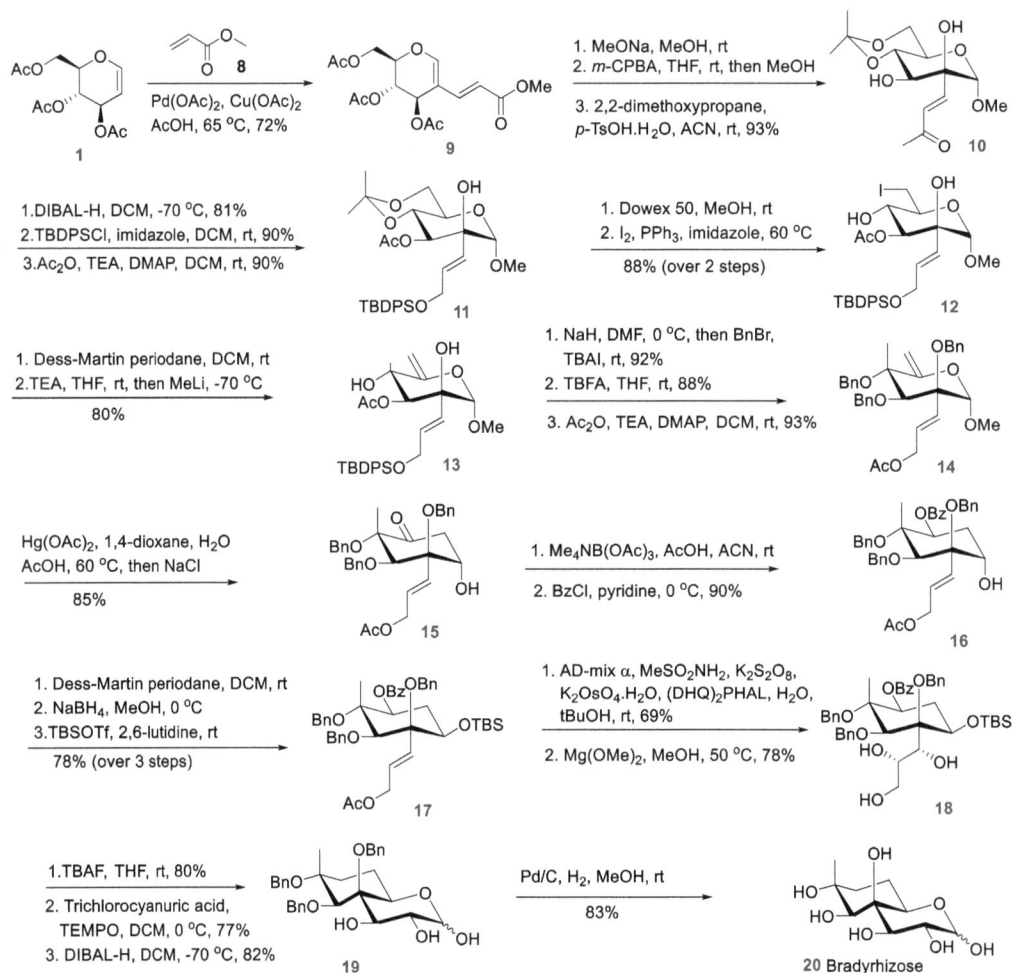

Scheme 2. Synthesis of Bradyrhizose.

Total synthesis of (+)-aspicilin

It is a naturally occurring molecule isolated from several lichens of the Lecanoraceae family. It shows cytotoxic activity toward A549, HeLa, and MCF7 cancer cells [15, 16]. Shaw and colleagues developed a synthetic strategy for (+)-Aspicilin **38** by using glucal **1a** as a chiral building block, as demonstrated in (Scheme 4) [17]. To develop appropriately protected glucal **30**, they started their synthesis by selectively protecting the primary hydroxy group with a bulky group like TBDPS and then protecting the secondary hydroxy groups with MOM.

Pyranose **31** was produced when glucal **30** was hydrated with LiBr in the presence of amberlite IR-120.

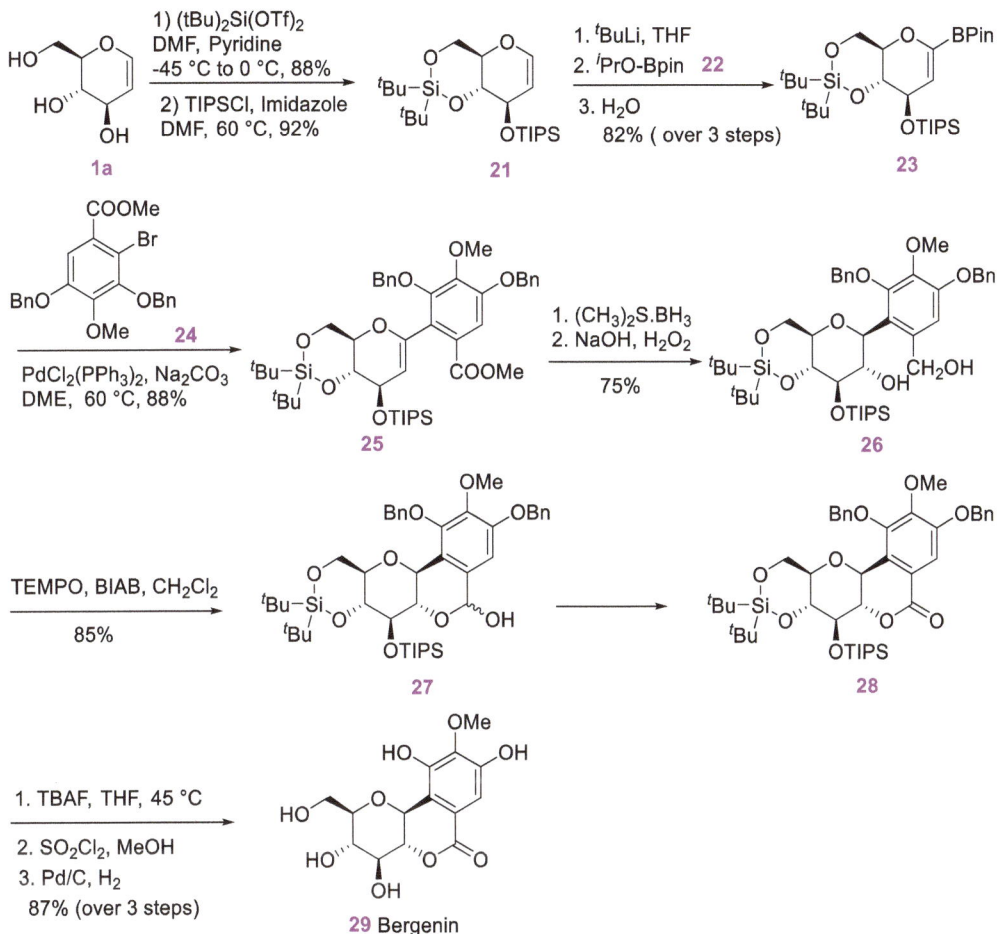

Scheme 3. Synthesis of Bergenin.

Olefin **32** was produced *via* the hydrolysis of the hemiacetal moiety on **31** and the subsequent Wittig olefination and MOM protection of the resultant hydroxy group. Desilylation of **32** with TBAH in THF resulted in the formation of **33**. The carboxylic acid analog **34** was produced by hydrolyzing the α,β-unsaturated ester obtained from the oxidation of the main hydroxy group *via* Swern oxidation using oxalyl chloride and the Wittig process. The essential first intermediate **36** was produced by esterification using a pre-formed hydroxy alkene **35**. Using Grubbs' generation for ring-closing metathesis produced macrolide **37**. The target (+)-aspicilin **38** was eventually obtained in an overall yield of 10% upon hydrogenation of the double bond by Pd/C & H$_2$ and removal of the MOM protective groups.

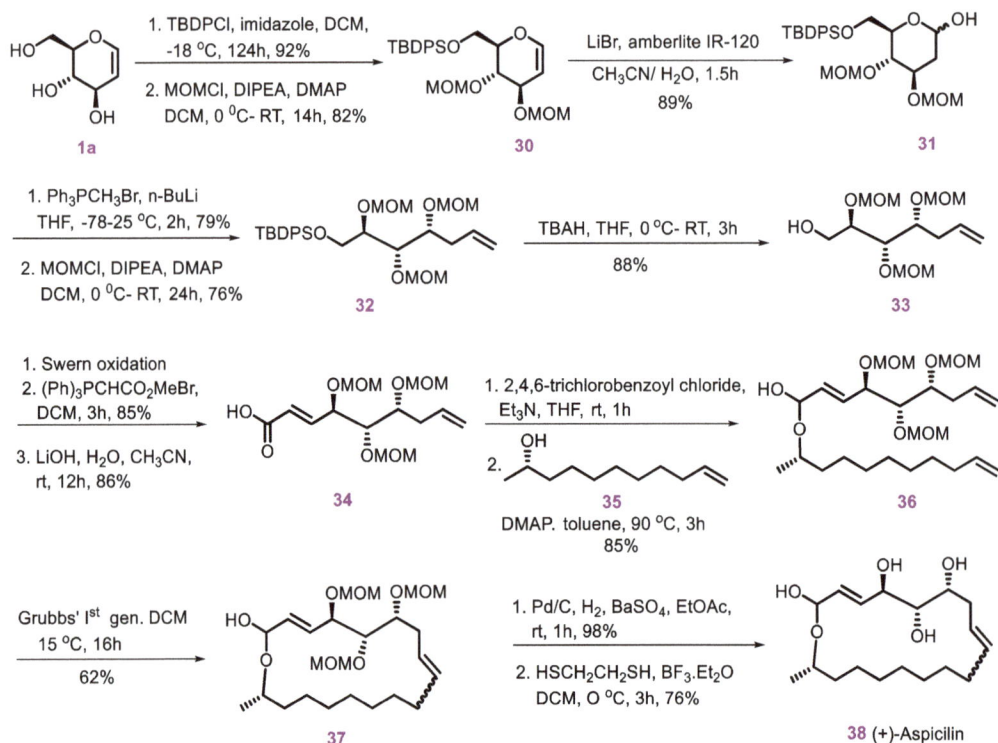

Scheme 4. Synthesis of (+)-aspicilin.

Synthesis of Reblastatin

Reblastatin **50** was initially discovered in the culture of *Streptomyces hygroscopicus* subsp. hygroscopicus SANK 61995 by Takatsu, and later isolated by Stead from *Streptomyces sp.* S6699. This compound exhibited significant anti-proliferative effects against human histicytic lymphoma U-937 cells (IC50 0.43 µg/mL or 0.78 µM) and demonstrated inhibitory properties against Oncostatin M (OSM- mediated pathways in HepG2 B6 cells (IC$_{50}$ 0.16 µM). These findings suggest potential applications in cancer treatment and rheumatoid arthritis [18 - 21]. A convergent synthesis of **50** in 19% total yield was reported by Yu and colleagues, as depicted in (Scheme **5**) [22]. They started their synthesis from commercially available 3,4,6-tri-*O*-acetyl-D-glucal **1** to build a C1-C7 fragment of **50**. Hemicetal is produced by heating glucal **1** in water, followed by hydrogenation to produce saturated hemiketal **39** in a diastereomeric combination. Wittig reaction of **40** gave an α, β-unsaturated ester **41;** after that, acetonide protection resulted in **42**. The methylation of the hydroxy group and acidic hydrolysis formed diol **43**. Oxidative cleavage of **43** yielded aldehyde **44** using sodium metaperiodate in a good amount. Compound **46** was synthesized through the coupling of **44** and alkyne **45** using a one-pot hydrozirconation/trans-metallation method. Compound **47** was obtained by protecting **46** with TBDMS, succeeded by hydrolysis and amidation employing BOC_2O/NH_4HCO_3 in dioxane and pyridine. When **47** is macrocyclized with copper iodide, macrolactam **48** is produced. Silyl deprotection and the amidation with trichloroacetyl isocyante resulted in compound **49**, which was then debenzylated using $AlCl_3$ and yielded the desired product reblastatin **50**.

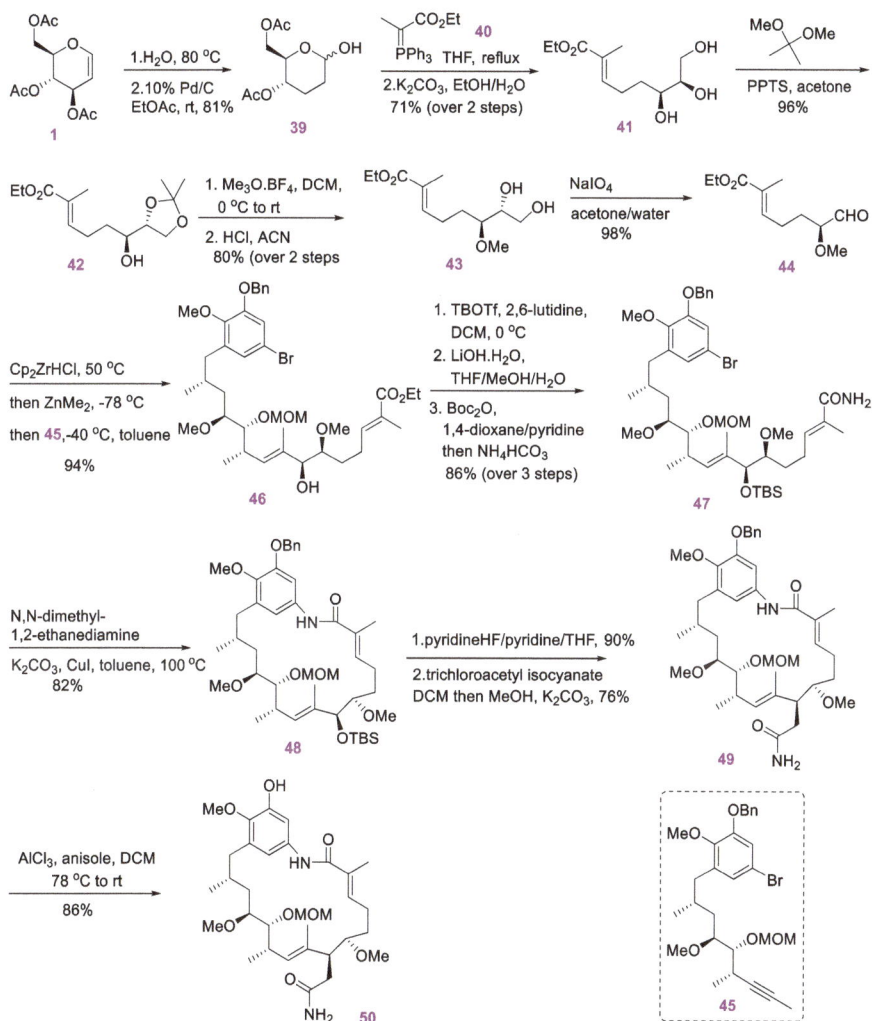

Scheme 5. Synthesis of reblastatin.

Synthesis of Cyclopropane Derivative of Spliceostatin A

The multimegadalton ribonucleoprotein complex known as spliceosome is responsible for catalyzing the joining of exons and the elimination of introns. Spliceosome inhibition is a rapidly expanding focus in the development of anticancer medications [23]. Spliceostatin A and its variants are among the many natural compounds that have the ability to effectively block spliceosomes [24]. Promising anticancer efficacy has been demonstrated by spliceostatin A and its variants. By adding a cyclopropane ring to the anomeric location of Spliceostatin A **59**, Ghosh *et al.* were able to synthesize the cyclopropane derivative, which has also demonstrated increased potency and stability [25]. Starting with glucal **1**, the group synthesized its corresponding benzylidine acetal **51** in two phases. Lactol was produced *via* the TBDMS protection of **51** and its subsequent reaction with NIS and Na₂S₂O₄. Lactol was oxidized to form the equivalent lactone **52**. Cyclopropane derivative **53** was produced by treating **52** with Petasis reagent and then performing Simmon-Smith cyclopropanation on it. **53** was converted using a traditional method into the olefin **54**, which was then reduced by using DIBAL-H to produce alcohol **55**. Vinyl iodide **56** was produced *via* oxidation of **55** and Takai olefination. Epoxide alcohol **57** was produced by PMB deprotection and TBHP epoxidation by VO(acac)₂. The targeted compound, which is a cyclopropane derivative of spliceostatin A **59**, was obtained in a yield of 29% as the sole product through Suzuki coupling of **57** with boronate **58** under Pd catalysis (Scheme **6**).

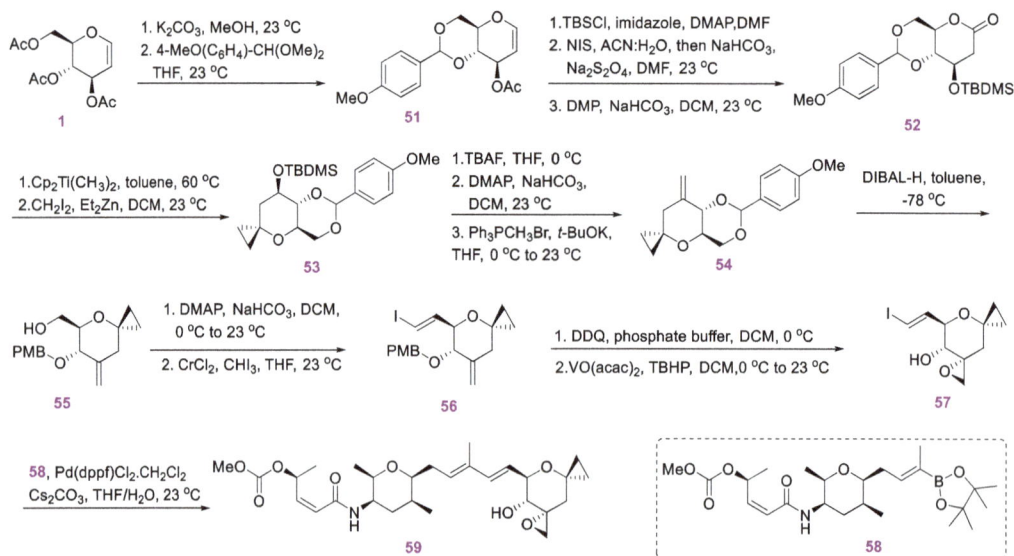

Scheme 6. Synthesis of biologically active cyclopropane derivative of spliceostatin A from 3,4,6-tri-*O*-acetyl-D-glucal.

Synthesis of Papulacandins A-D and its Derivatives

The natural antifungal substance that is extracted from Papulariasphaerosperma is called Papulacandins **A-D** and has shown anti-fungal activity [26, 27]. Denmark *et al.* [28]. synthesized Papulacandin D **70** in 31 steps using an anomeric-activated glucal derivative (Scheme **7**) with an overall yield of 9.2%. After doing a thorough examination, the author decided to use silanole **60** as the starting ingredient for the entire synthesis. To obtain the *C*-aryl glycoside **62**, silanol **60** was reacted with appropriately substituted aryl iodide **61** under Pd(II) catalysis. The oxidative spiroketalization process, in combination with DIBAL-H, produced spiroketal (**64a** and **64b**) as a mixture of α:β anomers. The molecule was generated *via* the acid-catalyzed conversion of the α -anomer into the β-anomer, TEOC protection of the C2 hydroxyl group, and TEOC protection and debenzylation of the aglycon benzyl groups **66**. The protected Papulacandin D **69** was created by TES deprotection and coupling of the fatty acid side-chain, which was synthesized from geraniol using **68**. Further, the deprotection of protecting groups in compound **69** resulted in Papulacandin D **70**.

Pieter *et al.* reported [29] the synthesis of papalucandin D derivatives that differ from papalucandin D by an acid side chain group (Scheme **8**). Using tri-*O*-acetyl D-glucal **1**, they first created anomeric activated glucal derivative **71**. Papulacandin D's aglycon moiety **72** was created by combining 3- and 5-hydroxy benzoic acid. The *C*-aryl glycoside **73**, which was produced *via* the coupling process between the aryl moiety **72** and silanol **71**, was catalyzed by palladium. To get α and β anomers, pyrolyl deprotection and oxidative spiroketalization were performed under standard conditions (**74a** and **74b**). In acidic conditions, the β anomers **74b** might easily be transformed into a stable α-anomer **74a**. The glycoside was obtained by debenzylation and MOM protection **75**. Therefore, to obtain the molecule, deprotection of silyl and reintroduction of cyclic silyl protecting group **77** was carried out. The subsequent coupling with distinct acids, succeeded by deprotection, resulted in the synthesis of derivatives of papalucandin D (**79a**, **79b**, **79c**, and **79d**).

Bergenin and the glucopyranoside moiety of Papulacandin D were also reported by Parkan *et al.*, as illustrated in (Scheme **9**) [30]. Boranated glucal **23** was synthesized from glucal **1** when appropriately reacted with substituted aryl iodide **80** to provide the matching glycoside **81**. Compound **81** undergoes epoxidation to give epoxide **82**. To obtain the required spiroketal, the benzylic hydroxyl's intramolecular assault opened the epoxide ring to get **74a**. The matching spiroketal moiety **84** was generated by debenzylation and desilylation followed by acetylation.

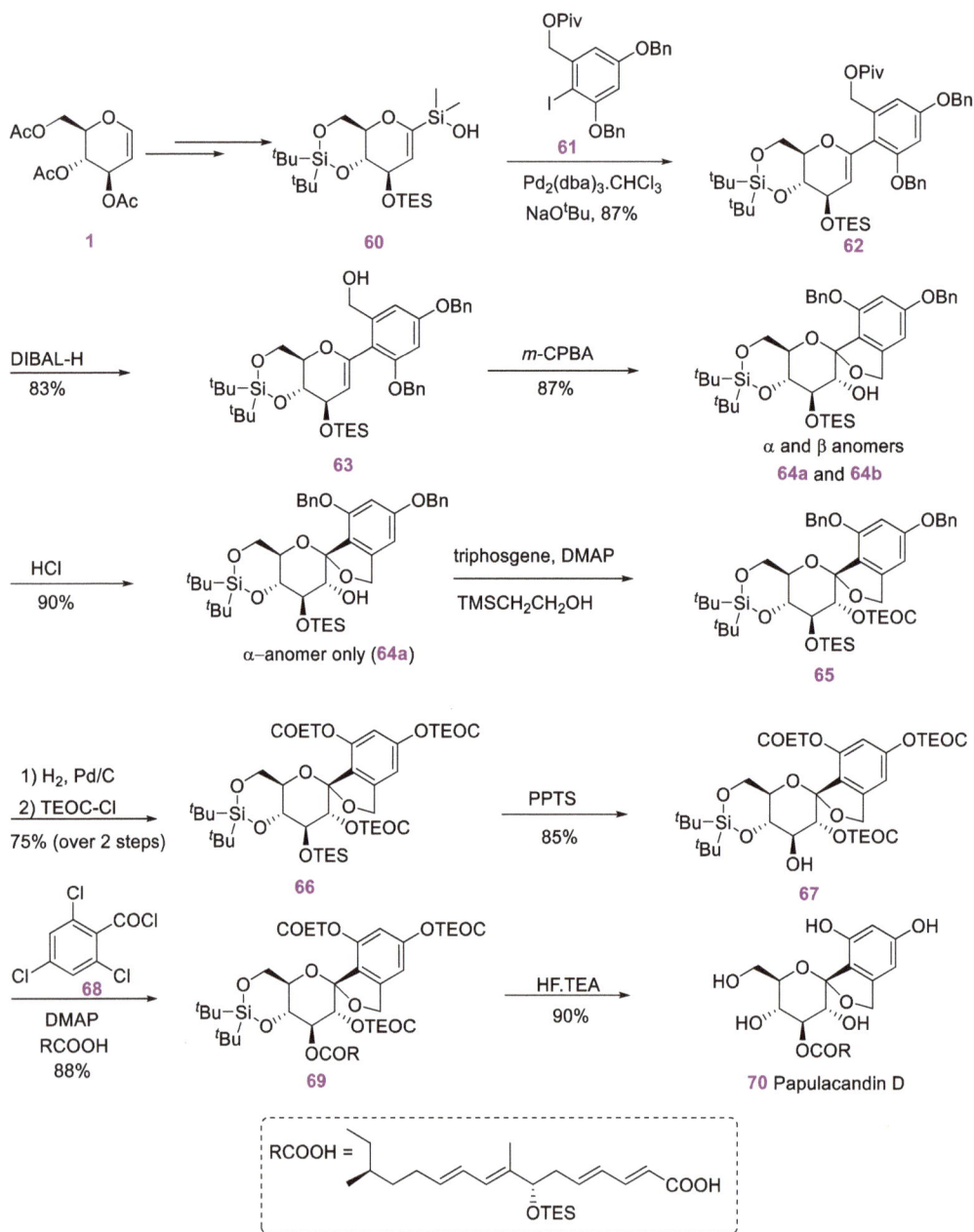

Scheme 7. Synthesis of Papulacandin D from tri-*O*-acetyl-D-glucal.

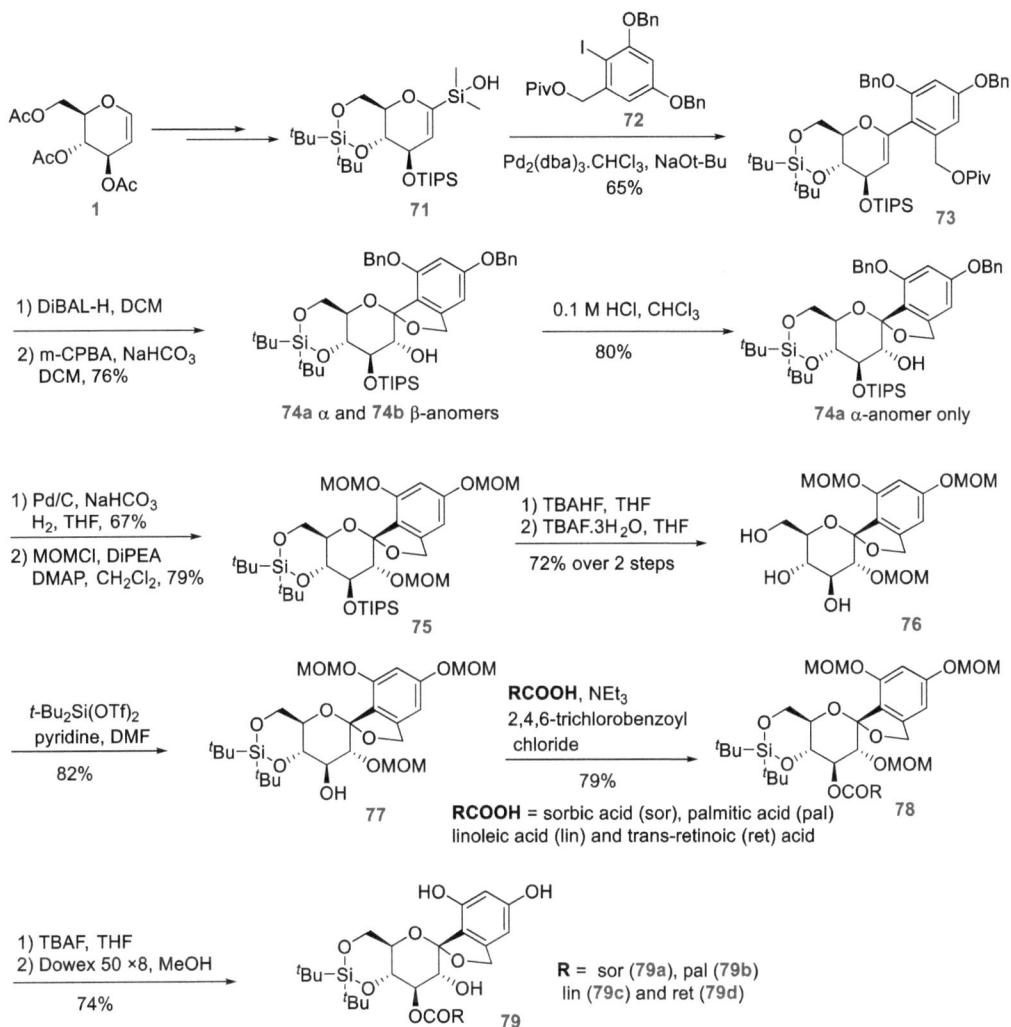

Scheme 8. Synthesis of Papulacandin D derivatives from tri-*O*-acetyl-D-glucal.

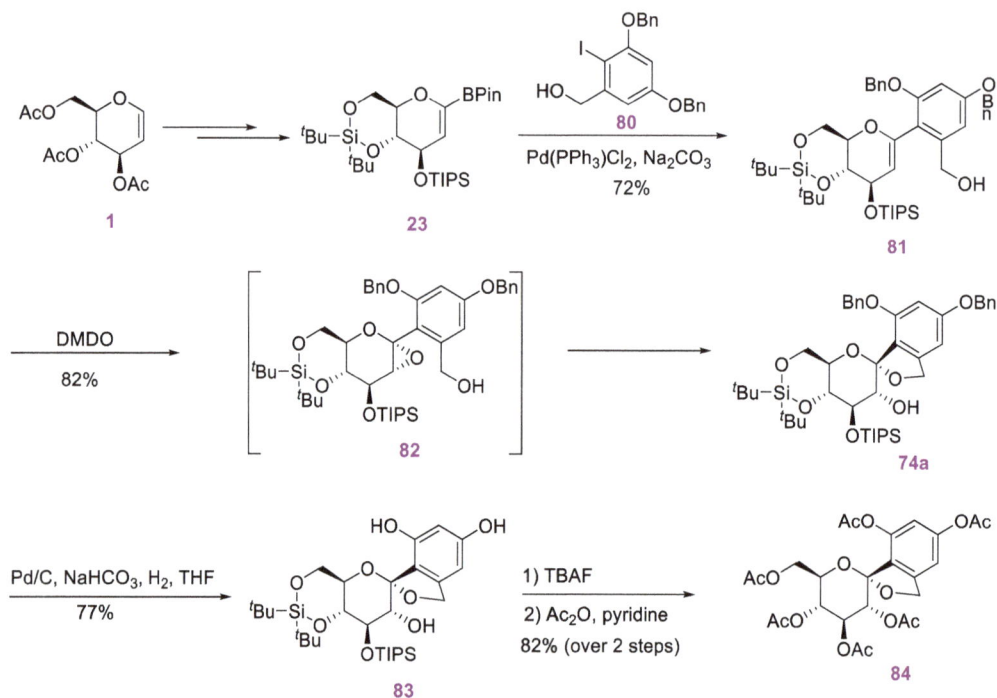

Scheme 9. Synthesis of glucopyranoside moiety papulacandin D derivatives from tri-*O*-acetyl-D-glucal.

Synthesis of (−)-Hyacinthacine A1

The natural product hyacinthacine A1 was isolated in less than 0.0005% from the bulbs of *Muscari armemiacum* (Hyacinthaceae) [31]. Hyacinthacine A1 is also a potential drug candidate for diabetes, viral infections, and cancer [32, 33]. 2-Nitroglycal, used as the precursor for the preparation of nitro-polyols, mediated by pyridine, was initially pioneered by Tang and his research team [34]. The produced substrates goes through protection-deprotection and Michael addition with acrylates, culminating in the production of (−)-hyacinthacine A1 **92**. The synthesis commences with glucal **85** as the starting material. In the first step, the endocyclic double bond is cleaved in the presence of concentrated HNO_3 and Ac_2O, yielding 67% of compound **86**. The addition of methyl acrylate to **86** leads to the formation of diastereomers (**87, 88**), which are subsequently reduced by $LiAlH_4$ in THF to produce alcohol **89** in 89% yield. Reacting diol with MsCl using triethylamine in DCM results in the protection of the primary OH group by the mesityl group, generating compound **90**. Subsequent treatment of compound **90** with Pd/H_2 produces compound **91** with an 85% yield. Finally, palladium-catalyzed debenzylation of compound **91** yields epi-(−)-hyacinthacine A1 **92** with an overall yield of 69% (Scheme **10**).

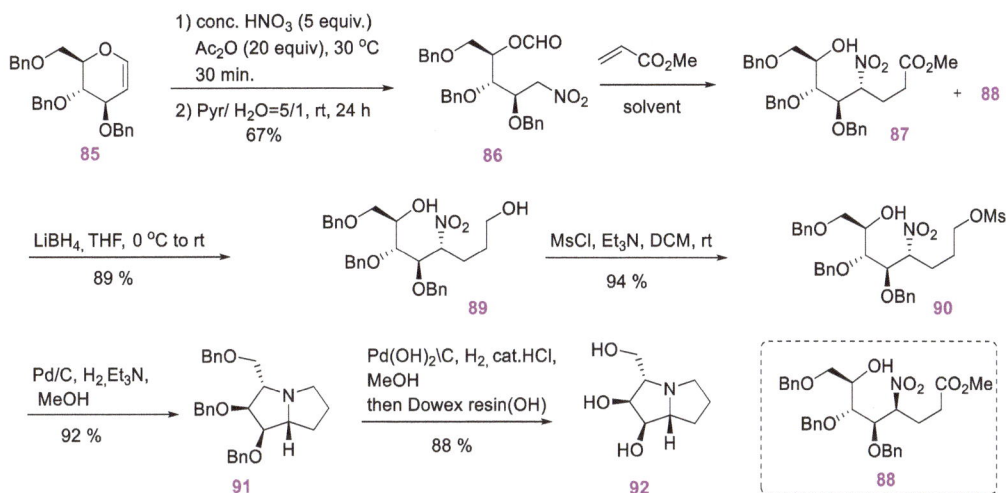

Scheme 10. Synthesis of (−)-Hyacinthacine A1.

Synthesis of Tricyclic flavonoid

Flavonoid **101**, structurally akin to vitexin and isovitexin, was extracted from oolong tea extract, showcasing robust anti-inflammatory properties [35, 36]. The synthesis was reported by Nakatsuka *et al.*, describing the transformation of 3,4,6-tri-*O*-benzyl-D-glucal **85** into a tricyclic flavonoid [37]. The synthetic pathway initiated with the epoxidation of **85** with DMDO succeeded by fluorination over four steps to yield glycosyl fluoride **94**. *C*-glycosylation with substituted acetophenone **95** and $BF_3 \cdot OEt_2$, subsequent methylation with methyl iodide, and deprotection (benzyl and PMB) yielded compound **97**. Cyclization of **97** by 1,10-azobis and Bu_3P gave compound **98**, followed by acylation and acyl group rearrangement under LDA conditions, resulting in the formation of compound **99**. Finally, cyclization and deprotection of benzyl using BCl_3 in DCM led to the production of tricyclic flavonoids, represented by compound **101** (Scheme **11**).

Sporiolide B and its derivatives

In 2004, Kobayashi reported the isolation of sporiolide B **102**, a 12-membered ring macrolide, from the culture broth of a fungus called *Cladosporium sp.* Spectrophotometric investigations reveal that the isolated macrolide structurally resembles the 3-*O*-methyl ether of pandangolide **103** [38]. Sporiolide B demonstrates antibacterial activity against *Micrococcus luteus* (MIC = 16.7 lg/mL) and cytotoxicity against L1210 cells (IC_{50} = 0.81 lg/mL) (Fig. **2**).

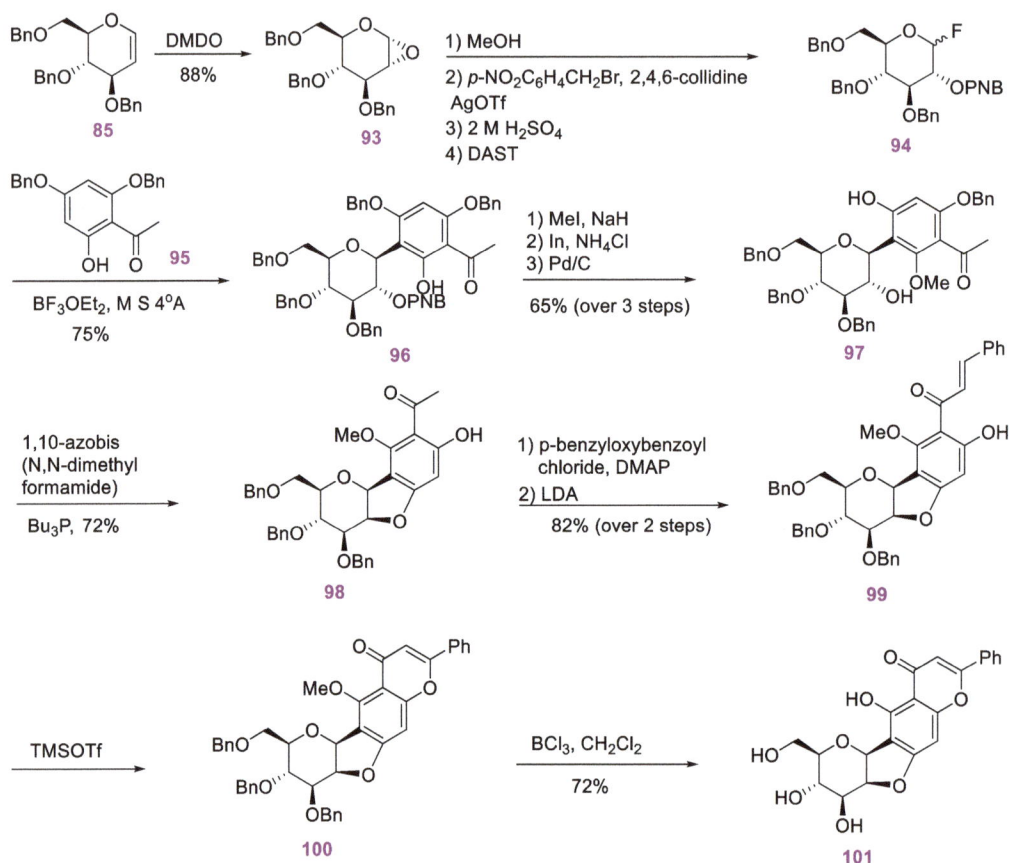

Scheme 11. Synthesis of tricyclic flavonoid.

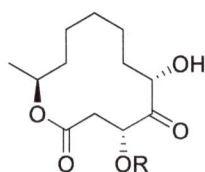

R=Me (Sporiolide B) **102**
R=H   (Pandangolide) **103**

Fig. (2). Sporiolide B and Pandangolide.

The complete synthesis of sporiolide B from 1,5-anhydro-2-deoxy-D-arabino-hex-1-enitol (D-glucal) was reported by Yuguo Du *et al.* [39] The synthesis was initiated from D-glucal **1a**. Using TBAF for desiccation and MeI for methylation, the glucal derivative **104** was produced. Lactone **105** was produced in a good yield by PCC oxidation of **104** at 45 °C in the presence of silica gel. Compound **105**, when in MeOH and treated with NaOMe, gives 83% methyl ester **106**. In the case of the Mitsunobu reaction, methyl ester produced C5 with inversion of stereochemistry followed by benzylation by using benzyl 2,2,2-trichloroacetimidate to obtain **107**. Compound **107** was hydrolyzed with LiOH and then esterified with (S)-6-hepten-2-ol by using Yamaguchi's conditions to obtain the diene analog **108** with protected hydroxy groups, which was subsequently treated with DDQ for PMB cleavage followed by Dess–Martin periodinane oxidation in DCM to get the crucial intermediate **109**. Using Grubbs catalyst [PhCH=RuCl(PCy$_3$)$_2$, 30 mol%], intramolecular ring closure metathesis (RCM) was carried out to create macrolactone **110** as an E:Z mixture (E/Z = 2:1). Ultimately, sporiloid B **102** was produced *via* palladium-catalyzed hydrogenation for the reduction of the double bond and simultaneous benzyl ether's cleavage (Scheme **12**).

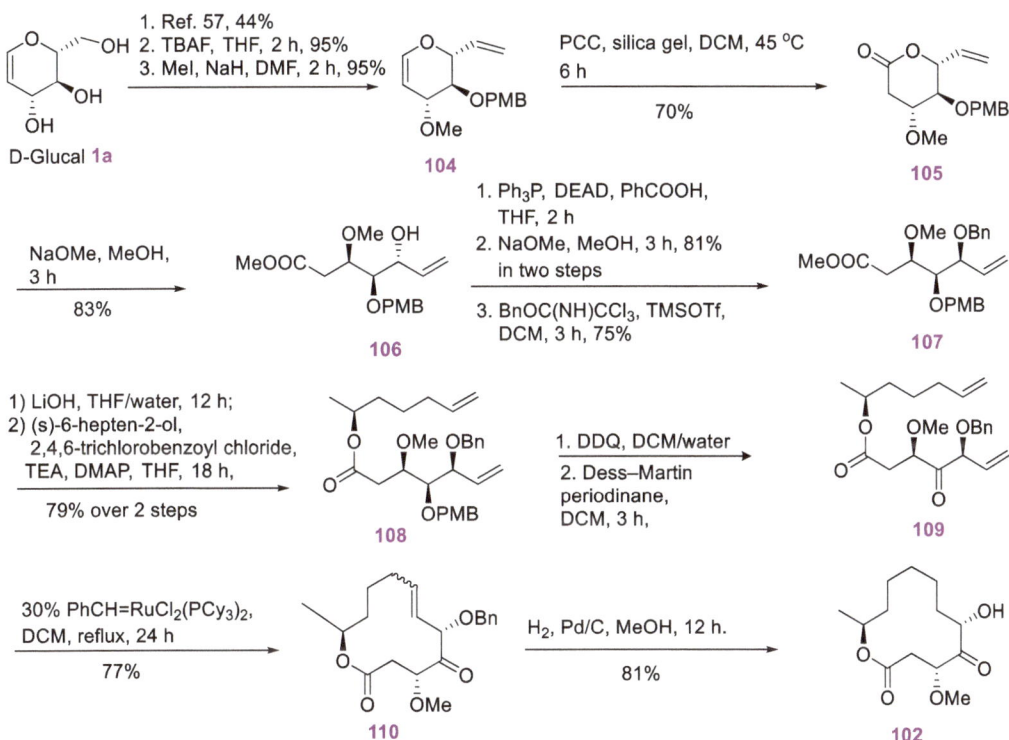

Scheme 12. Synthesis of Sporiolide B.

Synthesis of Cryptopyranmoscatone

In 2000, Cavalheiro and Yoshida reported the isolation of cryptopyranmoscatones extracted from the branch and stem bark of *Cryptocarya moschata*, a tree indigenous to the southeastern region of Brazil and belonging to the Lauraceae family. These compounds were identified through rigorous spectroscopic analyses

[40]. Notably, these 5,6-dihydro-alpha-pyrones are characterized by the presence of a styryl group attached to the C6 side chain. It is widely documented that styryl lactones exhibit considerable cytotoxicity against various human tumor cell lines [41]. Moreover, certain Cryptocarya pyrones have demonstrated potent inhibition of the G2 checkpoint, particularly in cells deficient in p53 function [42]. J. S. Yadav et al. achieved the first stereoselective total synthesis of cryptopyranmoscatone in 2010. The first step in the synthesis of cryptopyranmoscatones was the deacetylation of tri-*O*-acetyl-D-glucal **1**. This was followed by a selective silylation at position six and a treatment with MOM-Cl and Hunig's base in CH_2Cl_2 to produce a fully protected derivative of 3,4-di-*O*-MOM-6-*O*-silylated-D-glucal **111** [43]. After the silyl protecting group in **111** was removed with TBAF, primary alcohol **112** was produced in 70% yield. This alkene was then transformed into a D-glucal derivative alkene by further oxidation using IBX in $DMSO/CH_2Cl_2$, followed by a Wittig reaction. Pyridinium chlorochromate (PCC) in the presence of silica gel at 45°C then produced the key intermediate **113**. For the cross-metathesis reaction of olefin **113** with styrene **114**, Grubbs' second-generation carbene catalyst in benzene at 55 °C was utilized, resulting in lactone **115**. The lactol intermediate **116**, a mixture of two diastereomers, was easily obtained by treating lactone **115** and was refluxed with an excess of allylmagnesium bromide in THF. This process also underwent tandem stereoselective oxocarbenium cation formation, and the reduction of oxocarbenium cation was achieved by axial addition of Et_3SiH in the presence of TFA to yield glycoside **118**. From **115** (in >20:1 dr), *C*-glycoside **118** was produced in 68% overall yield. Previous reports served as the foundation for the supposition. Using Grubbs' second-generation catalyst and vinyl lactone **119** under reflux, the second cross-metathesis reaction of terminal alkene **118** produced the necessary lactone **120** with an 87% yield. Further deprotection of MOM groups was carried out with 4 N HCl in CH_3CN to synthesize cryptopyranmoscatone **121** (Scheme **13**).

Synthesis of Biologically Potent Molecule: Tetrahydroquinolines

Tetrahydroquinoline is one of the most important simple nitrogen heterocycles, being widespread in nature and present in a broad variety of pharmacologically active compounds [44]. Tetrahydroquinoline **130** was synthesized slightly from different intermediates **127**, as illustrated in (Scheme **14**) [45]. Performing a substitution reaction on iodomethyl acetate **123** with aniline proved challenging, as depicted in (Scheme **14**, resulting in the formation of carbaldehyde **126**. However, utilizing methyl glycoside **127**, derived from the opening of the cyclopropyl ring in **122**, along with *N*-alkylated anilines, yielded the desired intermediate **128**. This intermediate underwent acetolysis using $Ac_2O:AcOH:H_2SO_4$ to produce **129**, followed by intramolecular Friedel–Crafts alkylation using Sc(OTf)$_2$ in DCE, resulting in the formation of tetrahydroquinoline **130**.

Synthesis of Conduramine F4

Conduramines are aminocyclohexenetriols formally derived from conduritols, in which one of the hydroxyl groups is exchanged with an amino group [46a]. Many of these conduramines exhibit significant glycosidase inhibitory activity.

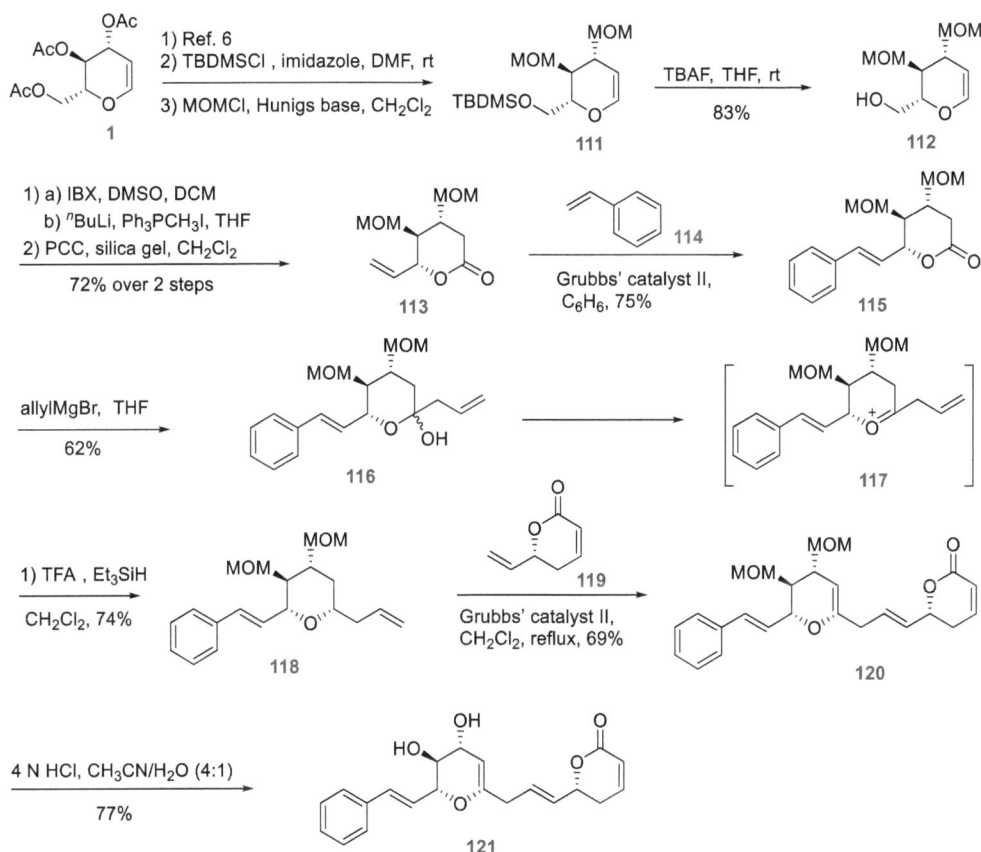

Scheme 13. Synthesis of cryptopyranmoscatone.

Ramesh and colleagues demonstrated an elegant synthesis of (-)-conduramine F4 **139** (Scheme **15**), starting from a common glucal **85** [46b]. Hemicetal was produced by iodoamination of glucal **85** using NiS and *p*-toluenesulfonamide in DCM and intramolecular migration of the amide by Et$_3$N, THF/H$_2$O. After hemiacetal underwent acetolysis, bis-acetate **133** was produced by ZnCl$_2$, which provided diol **134** by LAH as a crucial intermediary for reductive ring opening and the subsequent manipulation of protecting groups.

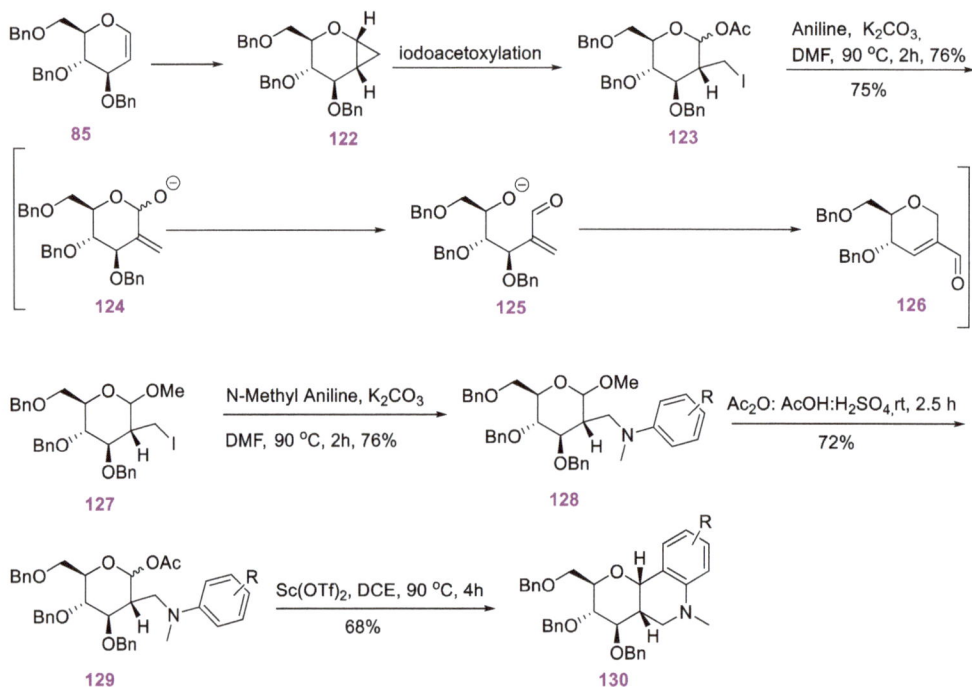

Scheme 14. Synthesis of tetrahydroquinolines starting from a protected glucal.

After removing the protective groups, (-)-conduramine F-4 was produced by Swern oxidation of the diol **135** to the dialdehyde **136**, which was followed by bis-Wittig olefination and Grubbs' generation catalyst-catalyzed olefin metathesis.

Synthesis of Aspergilide A

Liu and colleagues introduced a palladium-catalyzed decarboxylative glycosylation method for α-glycosylation of glycals, which was later expanded to accomplish the total synthesis of Aspergilide A **148** [47, 48]. The process commenced with the conversion of protected glucal **140** into the decarboxylative coupling product **141** using palladium acetate and DPPF, serving as the synthesis starting point. The reduction of the double bond in protected glucal **140** with Raney Ni and H_2 led to the formation of compound **142**.

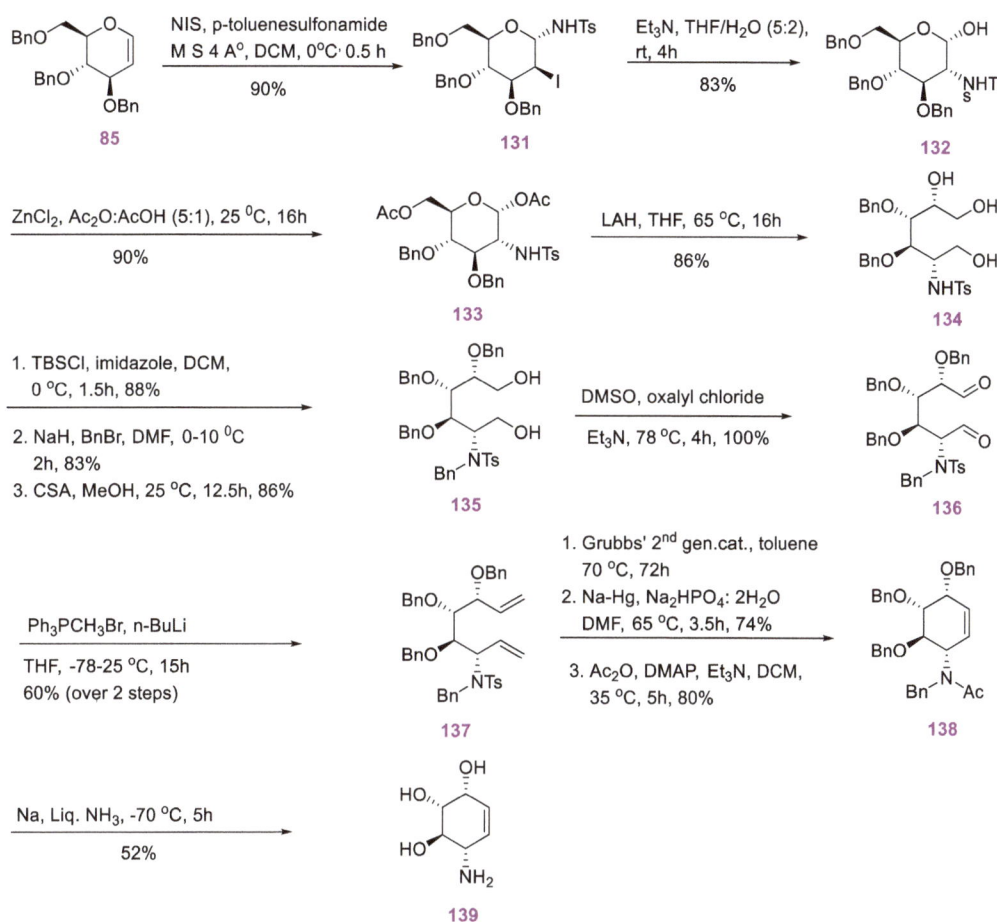

Scheme 15. Synthesis of conduramine F4.

Subsequent treatment of compound **142** with Tf$_2$O and DTBMP facilitated the generation of internal alkene **143** by eliminating the newly formed hydroxyl group. The cleavage of the acetal group with DIBAL-H liberated the primary hydroxyl group, followed by protection with TsCl and conversion into the nitrile group **144** through treatment with KCN in DMSO. The hydrolysis of the nitrile group under basic conditions yielded carboxylic acid **145**. Yamaguchi esterification was then employed with (*S*)-hept-6-en-2-ol to synthesize the precursor **146** for Grubb's cyclization. Treatment of compound **146** with Grubbs catalyst of the second-generation generated *Z*-alkene, which upon treatment with DDQ produced the PMB-deprotected compound **147**. Further isomerization of *Z*-alkene to *E*-alkene led to the formation of the target molecule **148** (Scheme **16**).

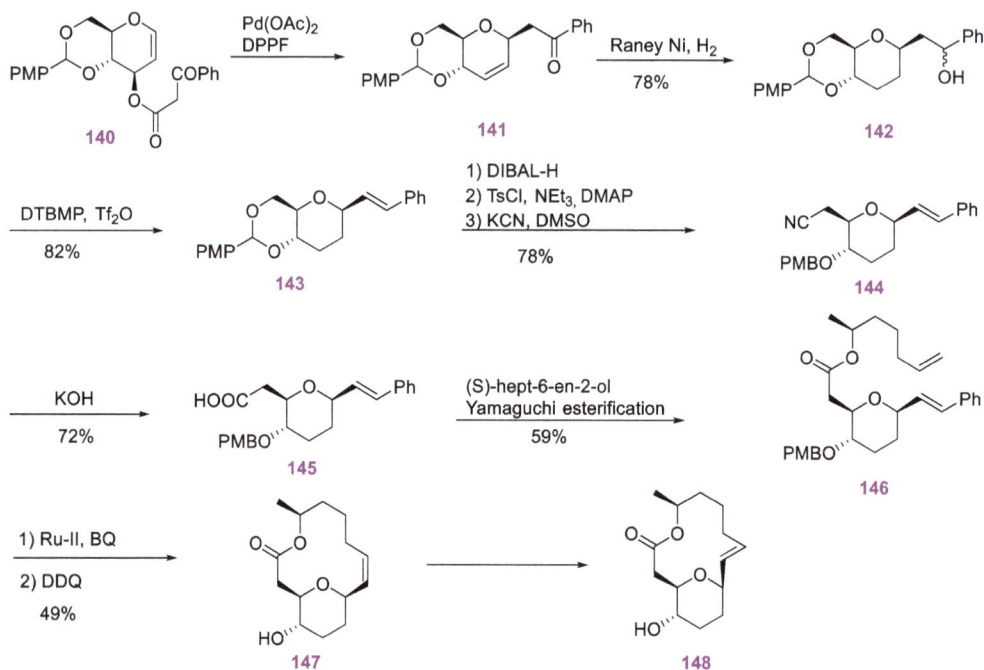

Scheme 16. Synthesis of Aspergilide A.

Synthesis of Vineomycinone B2

The preparation of vineomycinone B2 **158** methyl ester from appropriately protected glycal **149** was reported by Tiuset *et al.* [49].

The stereospecific *C*-aryl glycoside **151** is produced by the Pd-catalyzed coupling of iodo anthracene **150** with activated glucal, which is synthesized by interacting with *t*-BuLi and ZnCl₂. The double bond is then reduced with NaCNBH₄ to produce 2-deoxy aryl glycoside **152**. Compound **155** was obtained *via* selective ortho stannylation on the anthracene scaffold, which was subsequently coupled with bromo compound **154**. Molecule **157** was generated *via* methylation and oxidation. Additionally, it underwent deprotection using HCl in a single step, resulting in the molecule with the identical name **158** (Scheme 17).

Synthesis of Oxadecalin Core of Phomactin A

A unique natural substance, phomactin A, is obtained from the marine parasitic fungus Phomasp. It functions as a selective antagonist against 1-*O*-alkyl-2(R)-(acetyl glyceryl)-3-phosphorylcholine, the platelet-activating factor (PAF) [50, 51]. Numerous biological consequences, such as smooth muscle contraction, vascular permeability, hypotension, and platelet aggregation, have been linked to platelet-activating factor-mediated signaling. Additionally linked to the development of septic shock as well as inflammatory, cardiovascular, and pulmonary disorders is the platelet-activating factor [51, 52].

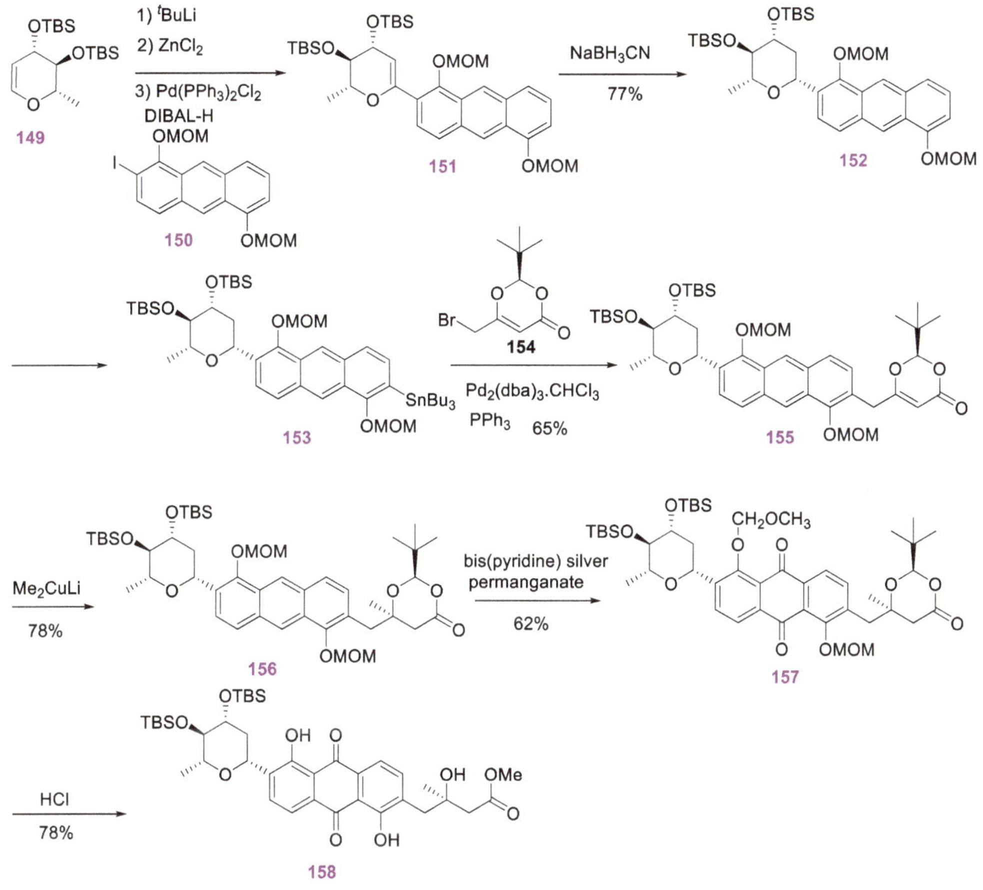

Scheme 17. Synthesis of Vineomycinone B2.

Phomactin A was achieved by Danishefsky *et al.* [53] in which Diels-Alder cycloaddition is considered a crucial step in its stereoselective synthesis of the oxadecalin core [54]. To synthesize the pyrone moiety **162**, the initial step involved a cycloaddition process utilizing substituted butadiene **159** and ethyl pyruvate **161**. Subsequently, oxidation of the pyrone moiety was performed, followed by iodination with molecular iodine and pyridine to yield the vinyl iodide moiety **164**. The keto group in molecule **164** was reduced through base treatment and application of Luche's conditions (CeCl₃, NaBH₄) to obtain the diol moiety **165**. Further modifications included TBDPS protection and ester group formation to yield compound **166**. The butadiene system underwent silyl protection, Stille coupling, and the introduction of the vinyl stannane **169**. Subsequent TBDMS protection using TBSCl and imidazole was followed by a Diels-Alder reaction with maleic anhydride **170** in CH_3CN, ultimately furnishing the desired product **171** (Scheme **18**).

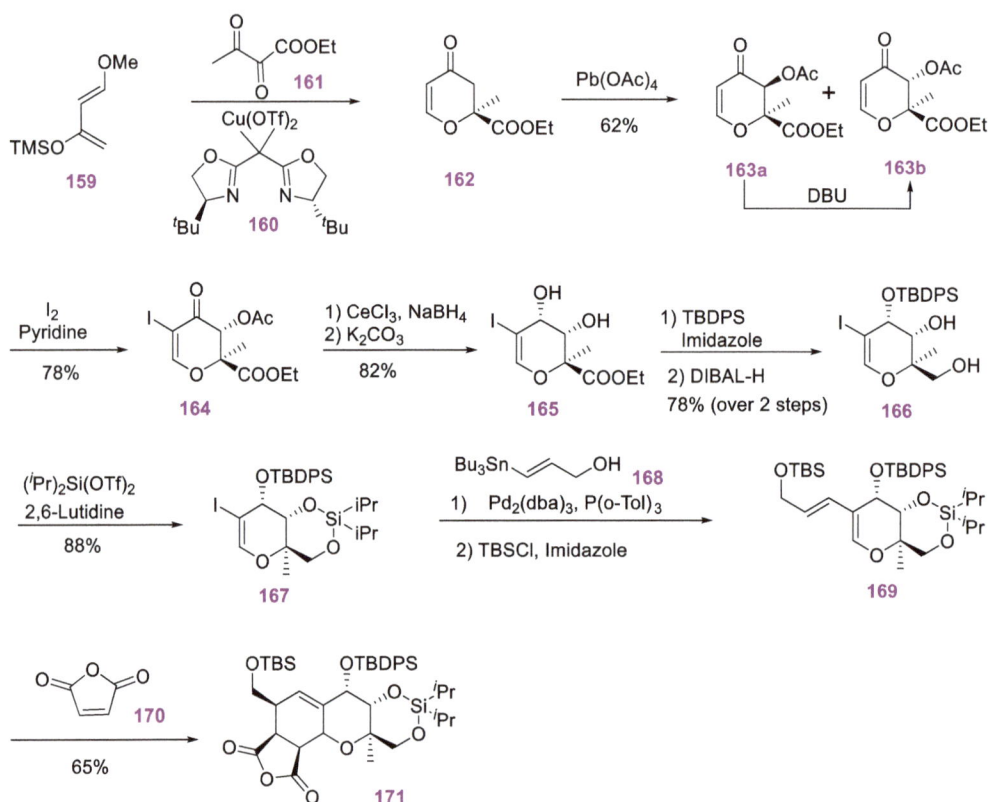

Scheme 18. Synthesis of oxadecalin core of Phomactin A.

Synthesis of Peptidomimetics

Using glucal **1a** as a common starting material, Giguère and colleagues synthesized several peptidomimetics with the goal of creating novel HIV protease inhibitors, as depicted in (Scheme **19**) [55]. Nucleophilic substitution of an azide provided **172** after the selective protection of the primary and secondary hydroxy groups of glucal **1a** as tosylate and acetate, respectively. When **172** were treated with *m*-CPBA in the presence of $BF_3 \cdot OEt_2$, α,β-unsaturated γ-lactone **173** was formed, which further underwent bicatalytic deacetylation and subsequent benzylation, leading to the generation of **174**. The addition of benzyl cuprate to **174** gives lactone **175**. Lactone **175** was produced by substituting a benzyl-protecting group for the acetate in **173** and then adding benzyl cuprate as a conjugate. A variety of gluconamides **176** were produced by heating lactone **175** with amines. These gluconamides were then functionalized to produce peptidomimetics, which showed possible antiproliferative properties.

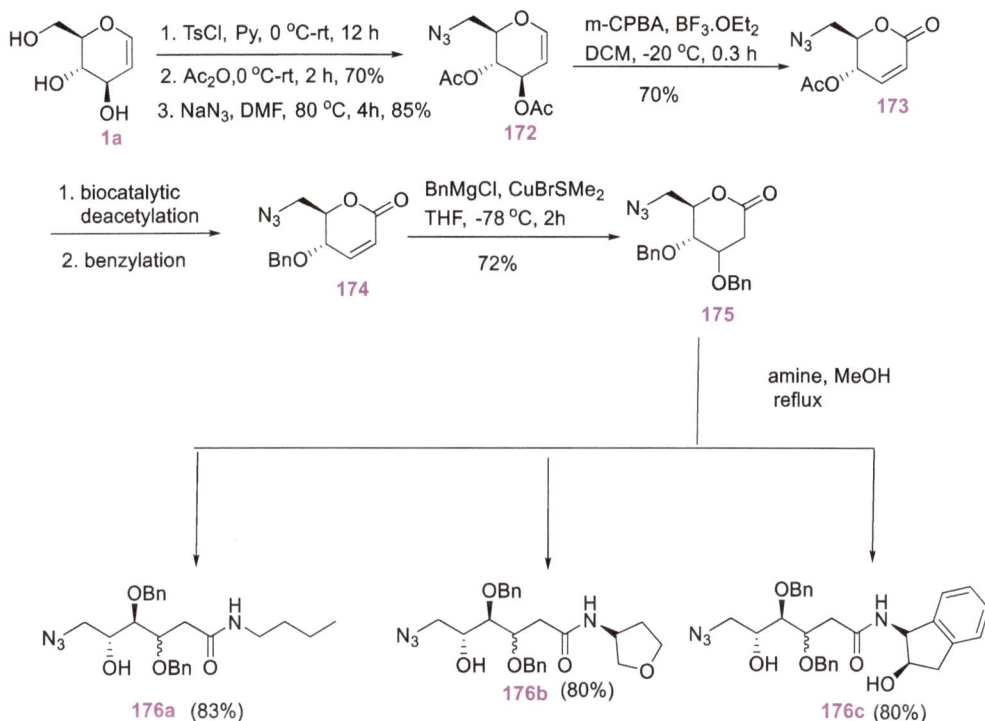

Scheme 19. Synthesis of peptidomimetics from glucal.

Synthesis of (−)-steviamine

Benzylated glucal **85** served as a pivotal precursor in the synthesis pathway of 1,8a-di-epi-(+)-steviamine **186a** and 2,3-diepi-(−)-steviamine **186b**, stereoisomers akin to the natural compound (−)-steviamine isolated from Stevia rebaudiana [56]. The plant has several medicinal claims, including anti-diabetic activity [57]. The synthetic route, outlined in (Scheme **20**) [58], commenced with the regioselective iodosulfonamidation of benzylated glucal **85**. Subsequent intramolecular substitution of the iodide, facilitated by migration of the NHTs group, resulted in the formation of glucopyranose **178**. Conversion of glucopyranose **178** into diol **179** involved the reduction of the hemiacetal using LiAlH$_4$, as several reagent systems were evaluated to ensure optimal reaction conditions (notably, milder reagents such as NaBH$_4$ proved ineffective). The primary hydroxy group in diol **179** was selectively protected with TBSCl, following a methodology analogous to the (+)-aspicilin synthesis described previously, leading to the formation of compound **180** (Scheme **20**). In the presence of PPh$_3$ and DEAD, intramolecular cyclization occurred, yielding pyrrolidine **181**. Subsequent removal of the silyl-protecting group, followed by oxidation of the resulting primary hydroxy group, produced 2-formyl pyrrolidine **180**, poised for a coupling reaction. Julia olefination of 2-formyl pyrrolidine **180** with a pre-formed sulfone **181** generated olefin pyrrolidine **182**, albeit with unintended epimerization at C-2.

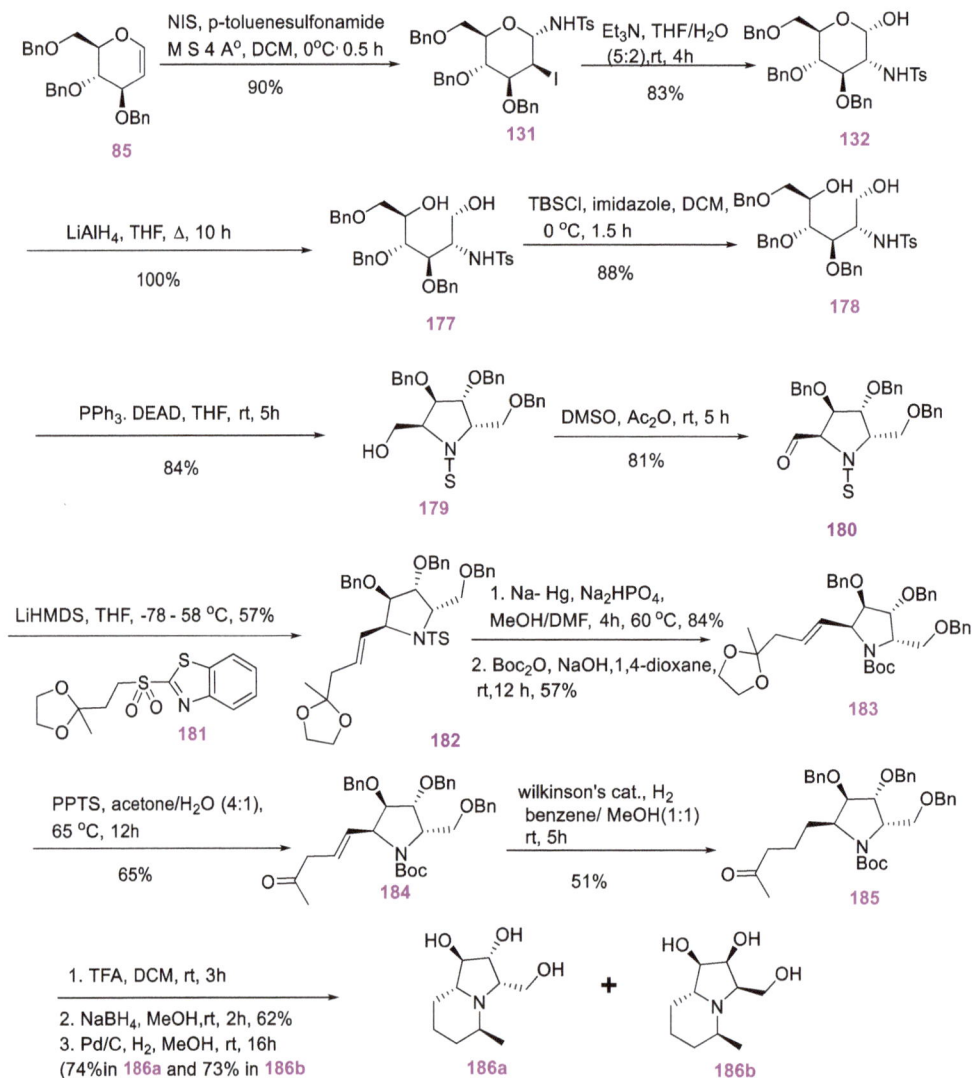

Scheme 20. Synthesis of (−)-steviamine **186b** from glucal.

Exchange of the Ts-protecting group on the nitrogen with Boc, followed by ketal deprotection, led to ketone **183**. Hydrogenation of the double bond in ketone **183** using Wilkinson's catalyst yielded saturated ketone **185**, which underwent Boc deprotection and intramolecular reductive amination. Subsequent removal of the benzyl-protecting groups *via* hydrogenolysis over Pd/C resulted in the synthesis completion, yielding **186a** and **186b**.

Synthesis of decytospolide A and B

Decytospolides A and B were isolated from the endophytic fungus *Cytospora* sp., an endophytic fungus from *Ilex canariensis,* by Zhang and co-workers [59] Decytospolide B 2 showed *in vitro* cytotoxic activity toward tumor cell lines A549 and QGY. Y. Fall and co-workers reported the synthesis of natural products (+)-decytospolides A **196** and B **195**, starting from glucal **1c**, as depicted in (Scheme **21**) [60].

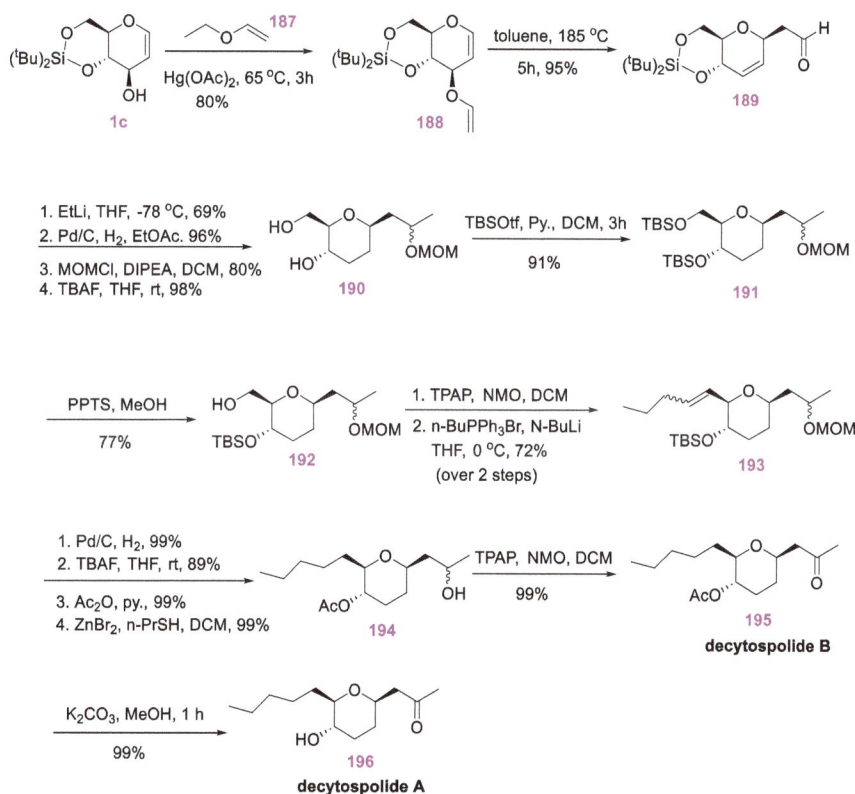

Scheme 21. Synthesis of natural products decytospolide A and B.

Protection of the hydroxy group as a vinyl ether **188**, followed by the classical Claisen rearrangement, led to the formation of aldehyde **189**. Subsequent 1,2-addition at the carbonyl center of the aldehyde with EtLi and reduction of the resulting double bond, followed by MOM protection of the resulting hydroxy group and desilylation of the existing protecting group, resulted in the formation of diol **190**. Re-masking of the hydroxy groups as TBS ethers with a subsequent selective desilylation of the primary TBS ether provided alcohol **192**. The oxidation of the alcohol into an aldehyde, followed by a Wittig olefination, yielded alkene **193** as a mixture of E/Z isomers in a 6:4 ratio. Hydrogenation of the mixture, along with protecting group manipulation, resulted in alcohol **194**, which was further oxidized to obtain decytospolide B **195**. Deacetylation of the protecting group led to the second natural product, decytospolide A **196**, as illustrated in (Scheme **21**).

Synthesis of Spliceostatin G

This family of natural products has been the subject of immense synthetic interest because they exhibit very potent cytotoxicity in representative human cancer cell lines. The cytotoxic properties of these natural products are related to their ability to inhibit spliceosomes [61 - 65]. They are isolated from the fermentation broth FERM BP-3421 of *Burkholderia* sp [61]. Ghosh and colleagues successfully accomplished the total synthesis of spliceostatin G **205**, utilizing acetylated glucal **1c** as a pivotal building block for the pyran fragment of the molecule (Scheme **22**) [66]. Firstly, glucal **188** was synthesized from *O*-allylation of glucal **1c**. Subsequent Claisen rearrangement followed by Wittig olefination facilitated the generation of 2,3-unsaturated pyranoside **197**. The removal of the silyl-protecting group and selective protection of the resulting primary hydroxy group using bulky 2,4,6-triisopropylbenzenesulfonyl chloride (TPSCl) yielded sulfonate **198**. Reduction of the sulfonate group, followed by oxidation of the alcohol and stereoselective Michael addition of a methyl group to the resulting α,β-unsaturated ketone, produced ketone **199**. The reduction of the ketone into an amine, achieved cleverly using a combination of NH_4OAc and $NaBH(OAc)_3$, followed by amidation, resulted in derivative **201**. This derivative underwent cross-metathesis with boronate **202** in the presence of Grubbs' 2nd generation catalyst, providing boronate **203**. A Suzuki–Miyaura cross-coupling reaction between boronate **203** and iodoacrylate as a coupling partner yielded a methyl ester, which was subsequently hydrolyzed and acetylated, ultimately furnishing spliceostatin G **205** in 68% yield over two steps.

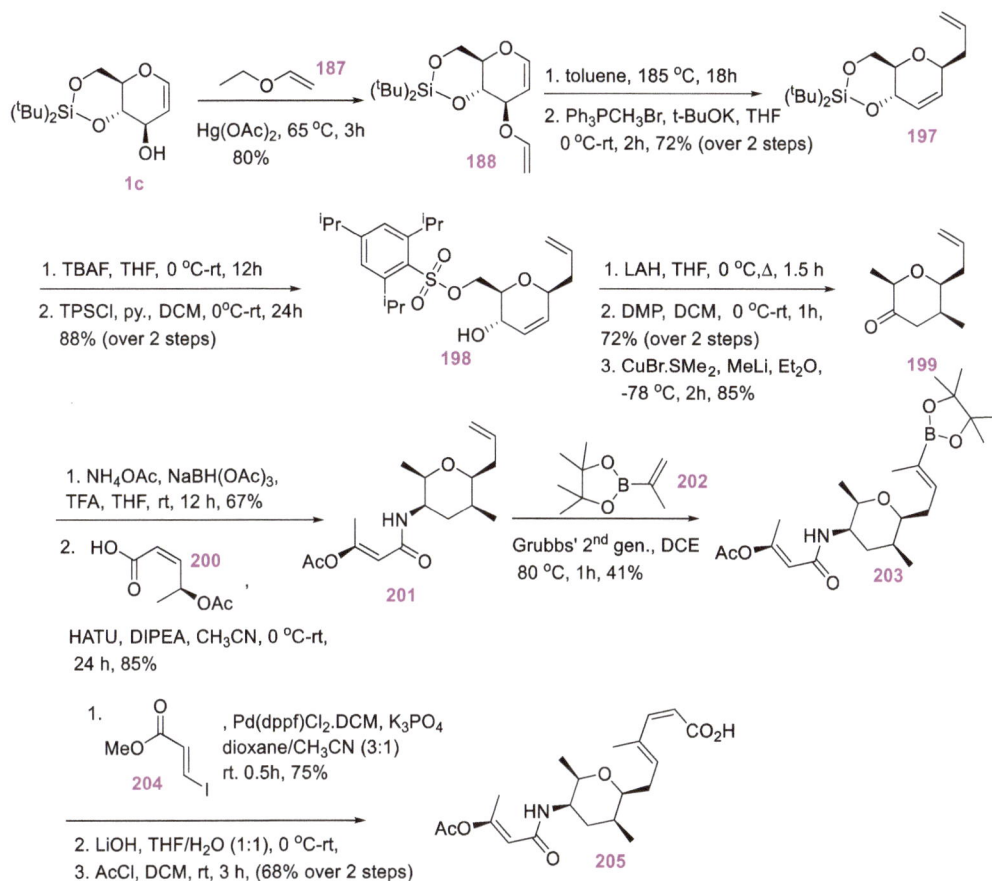

Scheme 22. Synthesis of spliceostatin G.

CONCLUSION

The comprehensive synthesis of natural products and pharmaceutically relevant compounds from glycals is an important field of organic chemistry research. Glycals are useful building blocks that can be used to create a variety of intricate molecular structures. Researchers now have access to a wide range of natural products and medicinally useful compounds with valuable biological activity. It is anticipated that additional research in this area will provide a new understanding of the synthesis processes as well as the possible uses of these molecules in the medical and associated disciplines.

ABBREVIATIONS

NaBH$_4$ Sodium borohydride

TBDMS   Tert-Butyldimethylsilyl-chloride

TBDPS	Tert-Butyldiphenylsilane
Cu(OAc)$_2$	Cupric acetate
TCCA	Trichloroisocyanuric acid
DIBAL-H	Diisobutylaluminium hydride
TIPS-Cl	Triisopropylsilyl chloride
t-BuLi	Tert-butyllithium
BH$_3$·Me$_2$S	Borane dimethylsulfide
LiBr	Lithium bromide
BOC$_2$O	Di-tert-butyl-dicarbonate
NH$_4$HCO$_3$	Ammonium hydrogen carbonate
PMB	P- methoxy benzyl
NIS	*N*-iodosuccinamide
Na$_2$S$_2$O$_4$	Sodium dithionite
VO(acac)$_2$	Vanadyl acetylacetonate
DIBAL	Diisobutylaluminium hydride
TEOC	2-(trimethylsilyl)ethoxycarbony
HNO$_3$	Nitric acid
DMDO	Dimethyldioxirne
LDA	Lithium diisopropylamide
TBAF	Tetrabutylammonium hydroxide
MeI	Methyl iodide
NaH	Sodium hydride
MeOH	Methanol
NaOMe	Sodium methoxide
LiOH	Lithium hydroxide
RCM	Ring closure metathesis
DDQ	2,3-dichloro-5,6-dicyano-1,4-benzoquinone
IBX	2-iodoxybenzoic acid
PCC	Pyridinium chlorochromate
THF	Tetrahydrofuran
CH$_3$CN	Acetonitrile
KCN	Potassium cyanide
DMSO	Dimethoxy sulfoxide

REFERENCES

[1]   (a) Fraser-Reid, B.; Radatus, B. 4,6-Di-O-acetyl-aldehydo-2,3-dideoxy-D-erythro-trans-hex-2-enose. Probable reason for the 'al' in Emil Fischer's triacetyl glucal. *J. Am. Chem. Soc.,* **1970**, *92*(17), 5288-5290.
[http://dx.doi.org/10.1021/ja00720a087]
(b) Fischer, E.; Zach, K. Sitzungsber. *Kgl. Preuss. Akad. Wiss,* **1913**, *27*, 311.

[2]   (a) Sakander, N.; Ahmed, A.; Zargar, I.A.; Mukherjee, D. Base-Mediated Transformation of Glycals to Their Corresponding Vinyl Iodides and Their Application in the Synthesis of C-3 Enofuranose and Bicyclic 3,4-Pyran-Fused Furanose. *J. Org. Chem.,* **2023**, *88*(13), 8300-8309.
[http://dx.doi.org/10.1021/acs.joc.3c00302] [PMID: 37315232]
(b) Ahmed, A.; Hussain, N.; Bhardwaj, M.; Chhalodia, A.K.; Kumar, A.; Mukherjee, D. Palladium catalysed carbonylation of 2-iodoglycals for the synthesis of C-2 carboxylic acids and aldehydes taking formic acid as a carbonyl source. *RSC Advances,* **2019**, *9*(39), 22227-22231.
[http://dx.doi.org/10.1039/C9RA03626A] [PMID: 35519467]
(c) Ahmed, A.; Rasool, F.; Singh, G.; Katoch, M.; Mukherjee, D. Synthesis and Conformational Analysis of 2-O-Silyl Protected Nucleosides from Unprotected Nucleobases and Sugar Epoxides. *Eur. J. Org. Chem.,* **2020**, *2020*(28), 4408-4416.
[http://dx.doi.org/10.1002/ejoc.202000650]
(d) Ahmed, A.; Sakander, N.; Rasool, F.; Hussain, N.; Mukherjee, D. Diastereoselective synthesis of glycopyrans 1,2-annulated with dioxazinanes from 1,2-anhydrosugars and *N* -substituted nitrones. *Org. Biomol. Chem.,* **2022**, *20*(7), 1436-1443.
[http://dx.doi.org/10.1039/D1OB02310A] [PMID: 35081611]
(e) Ahmed, A.; Sakander, N.; Mukherjee, D. *Chem. Select,* **2023**, *8*, e202300578.

[3]   (a) Kinfe, H.H. Versatility of glycals in synthetic organic chemistry: coupling reactions, diversity oriented synthesis and natural product synthesis. *Org. Biomol. Chem.,* **2019**, *17*(17), 4153-4182.
[http://dx.doi.org/10.1039/C9OB00343F] [PMID: 30893410]
(b) Hussain, N.; Hussain, A. Advances in Pd-catalyzed C–C bond formation in carbohydrates and their applications in the synthesis of natural products and medicinally relevant molecules. *RSC Advances,* **2021**, *11*(54), 34369-34391.
[http://dx.doi.org/10.1039/D1RA06351K] [PMID: 35497292]
(c) Hussain, N.; Ahmed, A.; Mukherjee, D. 2-Halo Glycals as "Synthon" for 2- *C* -Branched Sugar: Recent Advances and Applications in Organic Synthesis. *Asian J. Org. Chem.,* **2020**, *9*(6), 882-897.
[http://dx.doi.org/10.1002/ajoc.202000195]
(d) Ghouilem, J.; de Robichon, M.; Le Bideau, F.; Ferry, A.; Messaoudi, S. Emerging Organometallic Methods for the Synthesis of C-Branched (Hetero)aryl, Alkenyl, and Alkyl Glycosides: C–H Functionalization and Dual Photoredox Approaches. *Chemistry,* **2021**, *27*(2), 491-511.
[http://dx.doi.org/10.1002/chem.202003267] [PMID: 32813294]
(e) Lahiri, R.; Ansari, A.A.; Vankar, Y.D. Recent developments in design and synthesis of bicyclic azasugars, carbasugars and related molecules as glycosidase inhibitors. *Chem. Soc. Rev.,* **2013**, *42*(12), 5102-5118.
[http://dx.doi.org/10.1039/c3cs35525j] [PMID: 23535828]
(f) Jiang, N.; Wu, Z.; Dong, Y.; Xu, X.; Liu, X.; Zhang, J. Progress in the Synthesis of 2,3-unsaturated Glycosides. *Curr. Org. Chem.,* **2020**, *24*(2), 184-199.
[http://dx.doi.org/10.2174/1385272824666200130111142]
(g) Ahmed, A.; Mukherjee, D. Stereoselective Construction of Orthogonally Protected, N–O Interlinked Disaccharide Mimetics Using N-Substituted β-Aminooxy Donors. *J. Org. Chem.,* **2022**, *87*(8), 5125-5135.
[http://dx.doi.org/10.1021/acs.joc.1c03097] [PMID: 35357132]
(h) Hussain, N.; Rasool, F.; Khan, S.; Saleem, M.; Maheshwari, M. Advances in the Synthesis of Natural Products and Medicinally Relevant Molecules from Glycals. *ChemistrySelect,* **2022**, *7*(40), e202201873.
[http://dx.doi.org/10.1002/slct.202201873]

[4]   (a) Holt, D.J.; Barker, W.D.; Jenkins, P.R.; Davies, D.L.; Garratt, S.; Fawcett, J.; Russell, D.R.;

Ghosh, S. Ring-Closing Metathesis in Carbohydrate Annulation. *Angew. Chem. Int. Ed.,* **1998**, *37*(23), 3298-3300.
[http://dx.doi.org/10.1002/(SICI)1521-3773(19981217)37:23<3298::AID-ANIE3298>3.0.CO;2-O]
(b) Holt, D.J.; Barker, W.D.; Jenkins, P.R.; Panda, J.; Ghosh, S. Stereoselective preparation of enantiomerically pure annulated carbohydrates using ring-closing metathesis. *J. Org. Chem.,* **2000**, *65*(2), 482-493.
[http://dx.doi.org/10.1021/jo991392z] [PMID: 10813961]
(c) Bonnert, R.V.; Jenkins, P.R. The first example of a Robinson annulation on a carbohydrate derivative. *J. Chem. Soc. Chem. Commun.,* **1987**, (1), 6-7.
[http://dx.doi.org/10.1039/c39870000006]
(d) Wood, A.J.; Jenkins, P.R.; Fawcett, J.; Russell, D.R. The synthesis and X-ray crystal structure of a cyclopentaannulated sugar; the first example of an intramolecular aldol cyclopentaannulation in carbohydrate chemistry. *J. Chem. Soc. Chem. Commun.,* **1995**, (15), 1567-1568.
[http://dx.doi.org/10.1039/c39950001567]
(e) Wood, A.J.; Jenkins, P.R. *Tetrahedron Lett.,* **1997**, *38*, 1853-1856.
[http://dx.doi.org/10.1016/S0040-4039(97)00211-6]

[5]   (a) Xiao, M.; Wu, W.; Wei, L.; Jin, X.; Yao, X.; Xie, Z. Total synthesis of (−)-isatisine A *via* a biomimetic benzilic acid rearrangement. *Tetrahedron,* **2015**, *71*(22), 3705-3714.
[http://dx.doi.org/10.1016/j.tet.2014.09.028]
(b) Saidhareddy, P.; Ajay, S.; Shaw, A.K. 'Chiron' approach to the total synthesis of macrolide (+)-Aspicilin. *RSC Advances,* **2014**, *4*(9), 4253-4259.
[http://dx.doi.org/10.1039/C3RA45530K]
(c) Chen, Q.; Du, Y. Synthesis of sporiolide B from d-glucal. *Carbohydr. Res.,* **2007**, *342*(11), 1405-1411.
[http://dx.doi.org/10.1016/j.carres.2007.04.013] [PMID: 17517381]
(d) Tso, H-H.; Tsay, H. *Tetrahedron Lett.,* **1999**, *40*, 6869.
[http://dx.doi.org/10.1016/S0040-4039(99)01387-8]
(e) Saidhareddy, P.; Shaw, A.K. Glycal approach to the synthesis of macrolide (−)-A26771B. *RSC Advances,* **2015**, *5*(37), 29114-29120.
[http://dx.doi.org/10.1039/C4RA17084A]
(f) Sabitha, G.; Reddy, S.S.S.; Yadav, J.S.; Yadav, J.S. Total synthesis of cryptopyranmoscatone B1 from 3,4,6-tri-O-acetyl-d-glucal. *Tetrahedron Lett.,* **2010**, *51*(48), 6259-6261.
[http://dx.doi.org/10.1016/j.tetlet.2010.09.085]
(g) Tanimoto, H.; Saito, R.; Chida, N. Formal synthesis of (−)-morphine from d-glucal based on the cascade Claisen rearrangement. *Tetrahedron Lett.,* **2008**, *49*(2), 358-362.
[http://dx.doi.org/10.1016/j.tetlet.2007.11.037]
(h) Pazó, M.; Zúñiga, A.; Pérez, M.; Gómez, G.; Fall, Y. Total synthesis of (+)-decytospolides A and B. *Tetrahedron Lett.,* **2015**, *56*(24), 3774-3776.
[http://dx.doi.org/10.1016/j.tetlet.2015.04.055]

[6]   Chu, C.; Lu, Y.P.; Yin, L.; Hocher, B. The SGLT2 Inhibitor Empagliflozin Might Be a New Approach for the Prevention of Acute Kidney Injury. *Kidney Blood Press. Res.,* **2019**, *44*(2), 149-157.
[http://dx.doi.org/10.1159/000498963] [PMID: 30939483]

[7]   Jun, Y.; Kyoji, K.; Yasuhiro, T.; Quan, L.T.; Tatsuro, M.; Yingjie, C.; Shigetoshi, K. New Diarylheptanoids from the Rhizomes of *Dioscorea spongiosa* and Their Antiosteoporotic Activity. *Planta Med.,* **2004**, *70*(1), 54-58.
[http://dx.doi.org/10.1055/s-2004-815456] [PMID: 14765294]

[8]   Zúñiga, A.; Pérez, M.; Gándara, Z.; Fall, A.; Gómez, G.; Fall, Y. Synthesis of diospongin A, ent-diospongin A and C-5 epimer of diospongin B from tri-O-acetyl-D-glucal. *ARKIVOC,* **2015**, *2015*(7), 195-215.
[http://dx.doi.org/10.3998/ark.5550190.p009.191]

[9]   Lerouge, P.; Roche, P.; Faucher, C.; Maillet, F.; Truchet, G.; Promé, J.C.; Dénarié, J. Symbiotic host-specificity of *Rhizobium meliloti* is determined by a sulphated and acylated glucosamine oligosaccharide signal. *Nature,* **1990**, *344*(6268), 781-784.

[http://dx.doi.org/10.1038/344781a0] [PMID: 2330031]

[10] Li, W.; Silipo, A.; Molinaro, A.; Yu, B. Synthesis of bradyrhizose, a unique inositol-fused monosaccharide relevant to a Nod-factor independent nitrogen fixation. *Chem. Commun. (Camb.),* **2015**, *51*(32), 6964-6967.
[http://dx.doi.org/10.1039/C5CC00752F] [PMID: 25797311]

[11] Patel, DK; Patel, K; Kumar, R; Gadewar, M; Tahilyani, V **2012**; *2*, 163-167.

[12] Kim, H.S.; Lim, H.K.; Chung, M.W.; Kim, Y.C. Antihepatotoxic activity of bergenin, the major constituent of Mallotus japonicus, on carbon tetrachloride-intoxicated hepatocytes. *J. Ethnopharmacol.,* **2000**, *69*(1), 79-83.
[http://dx.doi.org/10.1016/S0378-8741(99)00137-3] [PMID: 10661887]

[13] De Abreu S. Lago IA, Souza GP, Piló-Veloso D, Duarte HA, C. Alcântara AF. *Org. Biomol. Chem.,* **2008**, *6*, 2713-2718.

[14] Parkan, K.; Pohl, R.; Kotora, M. Cross-coupling reaction of saccharide-based alkenyl boronic acids with aryl halides: the synthesis of bergenin. *Chemistry,* **2014**, *20*(15), 4414-4419.
[http://dx.doi.org/10.1002/chem.201304304] [PMID: 24590755]

[15] Hesse, O. J. Prakt. Chem. 1900, 62, 430; Hesse, O. *J. Prakt. Chem.,* **1904**, *70*, 449.
[http://dx.doi.org/10.1002/prac.19040700127]

[16] Reddy, C.R.; Rao, N.N.; Sujitha, P.; Kumar, C.G. *Eur. J. Org. Chem.,* **2012**, 1819.
[http://dx.doi.org/10.1002/ejoc.201101673]

[17] Saidhareddy, P.; Ajay, S.; Shaw, A.K. 'Chiron' approach to the total synthesis of macrolide (+)-Aspicilin. *RSC Advances,* **2014**, *4*(9), 4253-4259.
[http://dx.doi.org/10.1039/C3RA45530K]

[18] Wrona, I.E.; Gabarda, A.E.; Evano, G.; Panek, J.S. Total synthesis of reblastatin. *J. Am. Chem. Soc.,* **2005**, *127*(43), 15026-15027.
[http://dx.doi.org/10.1021/ja055384d] [PMID: 16248632]

[19] Stead, P.; Latif, S.; Blackaby, A.P.; Sidebottom, P.J.; Deakin, A.; Taylor, N.L.; Life, P.; Spaull, J.; Burrell, F.; Jones, R.; Lewis, J.; Davidson, I.; Mander, T. Discovery of Novel Ansamycins Possessing Potent Inhibitory Activity in a Cell-based Oncostatin M Signalling Assay. *J. Antibiot. (Tokyo),* **2000**, *53*(7), 657-663.
[http://dx.doi.org/10.7164/antibiotics.53.657]

[20] Takatsu, T.; Ohtsuki, M.; Muramatsu, A.; Enokita, R.; Kurakata, S.I. Reblastatin, a Novel Benzenoid Ansamycin-type Cell Cycle Inhibitor. *J. Antibiot. (Tokyo),* **2000**, *53*(11), 1310-1312.
[http://dx.doi.org/10.7164/antibiotics.53.1310]

[21] (a) Gorska, M.; Popowska, U.; Sielicka-Dudzin, A.; Kuban-Jankowska, A.; Sawczuk, W.; Knap, N.; Cicero, G.; Wozniak, F. Geldanamycin and its derivatives as Hsp90 inhibitors. *Front. Biosci.,* **2012**, *17*(7), 2269.
[http://dx.doi.org/10.2741/4050]
(b) Franke, J.; Eichner, S.; Zeilinger, C.; Kirschning, A. Targeting heat-shock-protein 90 (Hsp90) by natural products: geldanamycin, a show case in cancer therapy. *Nat. Prod. Rep.,* **2013**, *30*(10), 1299-1323.
[http://dx.doi.org/10.1039/c3np70012g] [PMID: 23934201]

[22] Bian, C.; Yan, R.; Yu, X. Total synthesis of reblastatin: convenient preparation of coupling partners and scaled assembly. *Tetrahedron,* **2014**, *70*(18), 2982-2991.
[http://dx.doi.org/10.1016/j.tet.2014.03.020]

[23] (a) Lee, S.C.W.; Abdel-Wahab, O. Therapeutic targeting of splicing in cancer. *Nat. Med.,* **2016**, *22*(9), 976-986.
[http://dx.doi.org/10.1038/nm.4165] [PMID: 27603132]
(b) Pal, S.; Gupta, R.; Kim, H.; Wickramasinghe, P.; Baubet, V.; Showe, L.C.; Dahmane, N.;

Davuluri, R.V. Alternative transcription exceeds alternative splicing in generating the transcriptome diversity of cerebellar development. *Genome Res.,* **2011**, *21*(8), 1260-1272.
[http://dx.doi.org/10.1101/gr.120535.111] [PMID: 21712398]

[24]   (a) van Alphen, R.J.; Wiemer, E A C.; Burger, H.; Eskens, F A L.M. The spliceosome as target for anticancer treatment. *Br. J. Cancer,* **2009**, *100*(2), 228-232.
[http://dx.doi.org/10.1038/sj.bjc.6604801] [PMID: 19034274]
(b) Hsu, T.Y.T.; Simon, L.M.; Neill, N.J.; Marcotte, R.; Sayad, A.; Bland, C.S.; Echeverria, G.V.; Sun, T.; Kurley, S.J.; Tyagi, S.; Karlin, K.L.; Dominguez-Vidaña, R.; Hartman, J.D.; Renwick, A.; Scorsone, K.; Bernardi, R.J.; Skinner, S.O.; Jain, A.; Orellana, M.; Lagisetti, C.; Golding, I.; Jung, S.Y.; Neilson, J.R.; Zhang, X.H.F.; Cooper, T.A.; Webb, T.R.; Neel, B.G.; Shaw, C.A.; Westbrook, T.F. The spliceosome is a therapeutic vulnerability in MYC-driven cancer. *Nature,* **2015**, *525*(7569), 384-388.
[http://dx.doi.org/10.1038/nature14985]
(c) Sidarovich, A.; Will, C.L.; Anokhina, M.M.; Ceballos, J.; Sievers, S.; Agafonov, D.E.; Samatov, T.; Bao, P.; Kastner, B.; Urlaub, H.; Waldmann, H.; Lührmann, R. Identification of a small molecule inhibitor that stalls splicing at an early step of spliceosome activation. *eLife,* **2017**, *6*, e23533.
[http://dx.doi.org/10.7554/eLife.23533] [PMID: 28300534]

[25]   Ghosh, A.K.; Reddy, G.C.; MacRae, A.J.; Jurica, M.S. Enantioselective Synthesis of Spliceostatin G and Evaluation of Bioactivity of Spliceostatin G and Its Methyl Ester. *Org. Lett.,* **2018**, *20*(1), 96-99.
[http://dx.doi.org/10.1021/acs.orglett.7b03456] [PMID: 29218995]

[26]   van der Kaaden, M.; Breukink, E.; Pieters, R.J. Synthesis and antifungal properties of papulacandin derivatives. *Beilstein J. Org. Chem.,* **2012**, *8*, 732-737.
[http://dx.doi.org/10.3762/bjoc.8.82] [PMID: 23015820]

[27]   (a) Traxler, P.; Gruner, J.; Auden, J.A.L. Papulacandins, a new family of antibiotics with antifungal activity. I. Fermentation, isolation, chemical and biological characterization of papulacandins A, B, C, D and E. *J. Antibiot. (Tokyo),* **1977**, *30*(4), 289-296.
[http://dx.doi.org/10.7164/antibiotics.30.289]
(b) Traxler, P.; Fritz, H.; Richter, W.J. [On the structure of papulacandin B, a new antibiotic with antifungal activity (author's transl)]. *Helv. Chim. Acta,* **1977**, *60*(2), 578-584.
[http://dx.doi.org/10.1002/hlca.19770600230] [PMID: 852996]
(c) Traxler, P.; Fritz, H.; Fuhrer, H.; Richter, W.J. Papulacandins, a new family of antibiotics with antifungal activity. Structures of papulacandins A, B, C and D. *J. Antibiot. (Tokyo),* **1980**, *33*(9), 967-978.
[http://dx.doi.org/10.7164/antibiotics.33.967]
(d) Traxler, P.; Gruner, J.; Nuesch, J. U. S. Patent **1981**-4278664, *4*, 278, 665.

[28]   Denmark, S.E.; Kobayashi, T.; Regens, C.S. Total synthesis of (+)-papulacandin D. *Tetrahedron,* **2010**, *66*(26), 4745-4759.
[http://dx.doi.org/10.1016/j.tet.2010.03.093] [PMID: 20711516]

[29]   Kaaden, M.; Breukink, E.; Pieters, J. *Beilstein J. Org. Chem.,* **2012**, *8*, 732-737.
[http://dx.doi.org/10.3762/bjoc.8.82] [PMID: 23015820]

[30]   Friesen, R.W.; Sturino, C.F. Stereoselective oxidative spiroketalization of a C-arylglucal derived from palladium-catalyzed coupling. Synthesis of the C-arylglucoside spiroketal nucleus of the papulacandins. *J. Org. Chem.,* **1990**, *55*(23), 5808-5810.
[http://dx.doi.org/10.1021/jo00310a004]

[31]   Asano, N.; Kuroi, H.; Ikeda, K.; Kizu, H.; Kameda, Y.; Kato, A.; Adachi, I.; Watson, A.A.; Nash, R.J.; Fleet, G.W.J. New polyhydroxylated pyrrolizidine alkaloids from Muscari armeniacum: structural determination and biological activity. *Tetrahedron Asymmetry,* **2000**, *11*(1), 1-8.
[http://dx.doi.org/10.1016/S0957-4166(99)00508-X]

[32]   Lillelund, V.H.; Jensen, H.H.; Liang, X.; Bols, M. Recent developments of transition-state analogue glycosidase inhibitors of non-natural product origin. *Chem. Rev.,* **2002**, *102*(2), 515-554.
[http://dx.doi.org/10.1021/cr000433k] [PMID: 11841253]

[33]   Pearson, M.S.M.; Mathe-Allainrnat, M.; Fargeas, V. Lebreton. *J. Eur. J. Org. Chem,* **2005**, *11*, 2159-2191.
[http://dx.doi.org/10.1002/ejoc.200400823]

[34]   Tang, S.; Xiong, D.C.; Jiang, S.; Ye, X.S. Nitro-polyols *via* Pyridine Promoted C=C Cleavage of 2-Nitroglycals. Application to the Synthesis of (−)-Hyacinthacine A1. *Org. Lett.,* **2016**, *18*(3), 568-571.
[http://dx.doi.org/10.1021/acs.orglett.5b03607] [PMID: 26794249]

[35]   Ishikura, Y.; Tuji, K.; Nukaya, H. WO2004-005296, **2004**.

[36]   Mahling, J.A.; Jung, K.H.; Schmidt, R.R. Glycosyl imidates, 69. Synthesis of flavone *C* -glycosides vitexin, isovitexin, and isoembigenin. *Liebigs Ann.,* **1995**, *1995*(3), 461-466.
[http://dx.doi.org/10.1002/jlac.199519950362]

[37]   Nakatsuka, T.; Tomimori, Y.; Fukuda, Y.; Nukaya, H. First total synthesis of structurally unique flavonoids and their strong anti-inflammatory effect. *Bioorg. Med. Chem. Lett.,* **2004**, *14*(12), 3201-3203.
[http://dx.doi.org/10.1016/j.bmcl.2004.03.108] [PMID: 15149675]

[38]   Gesner, S.; Cohen, N.; Ilan, M.; Yarden, O.; Carmeli, S. Pandangolide 1a, a metabolite of the sponge-associated fungus *Cladosporium* sp., and the absolute stereochemistry of pandangolide 1 and iso-cladospolide B. *J. Nat. Prod.,* **2005**, *68*(9), 1350-1353.
[http://dx.doi.org/10.1021/np0501583] [PMID: 16180812]

[39]   Du, Y.; Chen, Q.; Linhardt, R.J. The first total synthesis of sporiolide A. *J. Org. Chem.,* **2006**, *71*(22), 8446-8451.
[http://dx.doi.org/10.1021/jo0615504] [PMID: 17064018]

[40]   Cavalheiro, A.J.; Yoshida, M. 6-[ω-arylalkenyl]-5,6-dihydro-α-pyrones from *Cryptocarya moschata* (Lauraceae). *Phytochemistry,* **2000**, *53*(7), 811-819.
[http://dx.doi.org/10.1016/S0031-9422(99)00532-4] [PMID: 10783987]

[41]   (a) Fang, X.P.; Anderson, J.E.; Chang, C.J.; McLaughlin, J.L.; Fanwick, P.E. Two new styryl lactones, 9-deoxygoniopypyrone and 7-epi-goniofufurone, from *Goniothalamus giganteus. J. Nat. Prod.,* **1991**, *54*(4), 1034-1043.
[http://dx.doi.org/10.1021/np50076a017] [PMID: 1791471]
(b) Fang, X.; Anderson, J.E.; Chang, C.; McLaughlin, J.L. Three new bioactive styryllactones from goniothalamus giganteus (Annonaceae). *Tetrahedron,* **1991**, *47*(47), 9751-9758.
[http://dx.doi.org/10.1016/S0040-4020(01)80715-8]
(c) Tsubuki, M.; Kanai, K.; Nagase, H.; Honda, T. Stereocontrolled syntheses of novel styryl lactones, (+)-goniodiol, (+)-goniotriol, (+)-8-acetylgoniotriol, (+)-goniofufurone, (+)-9-deoxygoniopypyrone, (+)-goniopypyrone, and (+)-altholactone from common intermediates and cytotoxicity of their congeners. *Tetrahedron,* **1999**, *55*(9), 2493-2514.
[http://dx.doi.org/10.1016/S0040-4020(99)00023-X]

[42]   Sturgeon, C.M.; Cinel, B.; Diaz-Marrero, A.R.; McHardy, L.M.; Ngo, M.; Andersen, R.J.; Roberge, M. Chemother. C. *Pharmacol.,* **2008**, *61*, 407-413.

[43]   Scholl, M.; Ding, S.; Lee, C.W.; Grubbs, R.H. Synthesis and activity of a new generation of ruthenium-based olefin metathesis catalysts coordinated with 1,3-dimesityl-4,5-dihydroimidaz-l-2-ylidene ligands. *Org. Lett.,* **1999**, *1*(6), 953-956.
[http://dx.doi.org/10.1021/ol990909q] [PMID: 10823227]

[44]   Muthukrishnan, I.; Sridharan, V.; Menéndez, J.C. Progress in the Chemistry of Tetrahydroquinolines. *Chem. Rev.,* **2019**, *119*(8), 5057-5191.
[http://dx.doi.org/10.1021/acs.chemrev.8b00567] [PMID: 30963764]

[45]   Moshapo, P.T.; Kinfe, H.H. *Synthesis,* **2015**, *47*, 3673.
[http://dx.doi.org/10.1055/s-0035-1560174]

[46]   (a) Rajender, A.; Rao, B.V. Stereoselective synthesis of (−)-conduramine C-1 and (−)-conduramine D-

1. *Tetrahedron Lett.,* **2013**, *54*(19), 2329-2331.
[http://dx.doi.org/10.1016/j.tetlet.2013.02.046]
(b) Harit, V.K.; Ramesh, N.G. A Chiron Approach to Diversity-Oriented Synthesis of Aminocyclitols,
(−)-Conduramine F-4 and Polyhydroxyaminoazepanes from a Common Precursor. *J. Org. Chem.,*
2016, *81*(23), 11574-11586.
[http://dx.doi.org/10.1021/acs.joc.6b01790] [PMID: 27806198]

[47]   For the synthesis of aspergillide A, see: a) Sharma, G. V. M.; Manohar, V. *Tetrahedron Asymmetry*
2012, *23*, 252–263. b) Andrea, Z.; Manuel, P.; Maria, G.; Generosa, G.; Yagamare, F. *Synthesis* **2011**,
3301–3306. c) Izuchi, Y.; Kanomata, N.; Koshino, H.; Hongo, Y.; Nakata, T.; Takahashi, S.
Tetrahedron Asymmetry **2011**, *22*, 246–251. d) Nagasawa, T.; Nukada, T.; Kuwahara, S. *Tetrahedron*
2011, *67*, 2882–2888. e) Kanematsu, M.; Yoshida, M.; Shishido, K. *Angew. Chem. Int. Ed.* **2011**, *50*,
2618–2620. f) Fuwa, H.; Yamaguchi, H.; Sasaki, M. *Tetrahedron* **2010**, *66*, 7492–7503. g) Sabitha,
G.; Reddy, D. V.; Rao, A. S.; Yadav, J. S. *Tetrahedron Lett.* **2010**, *51*, 4195–4198. h) Nagasawa, T.;
Kuwahara, S. Tetrahedron Lett. 2010, 51, 875–877. i) Díaz-Oltra, S.; Angulo-Pachón, C. A.; Murga,
J.; Carda, M.; Marco, J. A. *J. Org. Chem.* **2010**, *75,* 1775–1778. j) Fuwa, H.; Yamaguchi, H.; Sasaki,
M. *Org. Lett.,* **2010**, *12*, 1848-1851.

[48]   For the isolation and structure revision of aspergillide A, see: a) Kito, K.; Ookura, R.; Yoshida, S.;
Namikoshi, M.; Ooi, T.; Kusumi, T. *Org. Lett.* **2008**, *10*, 225. b) Ookura, R.; Kito, K.; Saito, Y.;
Kusumi, T.; Ooi, T. *Chem. Lett.,* **2009**, *38*, 384.

[49]   Tius, M.A.; Gu, X.Q.; Gomez-Galeno, J. Convergent synthesis of vineomycinone B2 methyl ester. *J.*
Am. Chem. Soc., **1990**, *112*(22), 8188-8189.
[http://dx.doi.org/10.1021/ja00178a065]

[50]   Sugano, M.; Sato, A.; Iijima, Y.; Oshima, T.; Furuya, K.; Kuwano, H.; Hata, T.; Hanzawa, H.
Phomactin A; a novel PAF antagonist from a marine fungus Phoma sp. *J. Am. Chem. Soc.,* **1991**,
113(14), 5463-5464.
[http://dx.doi.org/10.1021/ja00014a053]

[51]   (a) Koltai, M.; Braquet, P.G. Platelet-activating factor antagonists. *Clin. Rev. Allergy,* **1995**, *12*(4),
361-380.
[http://dx.doi.org/10.1007/BF02802300] [PMID: 7743462]
(b) Goldstein, R.E.; Feuerstein, G.Z.; Bradley, L.M.; Stambouly, J.J.; Laurindo, F.R.M.; Davenport,
N.J. Cardiovascular effects of platelet-activating factor. *Lipids,* **1991**, *26*(12Part2), 1250-1256.
[http://dx.doi.org/10.1007/BF02536542] [PMID: 1819712]
(c) Rabinovici, R.; Yue, T.L.; Feuerstein, G. Platelet-activating factor in cardiovascular stress
situations. *Lipids,* **1991**, *26*(12Part2), 1257-1263.
[http://dx.doi.org/10.1007/BF02536543] [PMID: 1819735]

[52]   (a) Heuer, H.O. Involvement of platelet-activating factor (PAF) in septic shock and priming as
indicated by the effect of hetrazepinoic PAF antagonists. *Lipids,* **1991**, *26*(12Part2), 1369-1373.
[http://dx.doi.org/10.1007/BF02536569] [PMID: 1819735]
(b) Bazen, N.G.; Squinto, S.P.; Branquet, P.; Panetta, T.; Marchelselli, V.L. *Lipids,* **1991**, *26*, 1236-
1242.
[http://dx.doi.org/10.1007/BF02536539] [PMID: 1668121]
(c) Uchiyama, S.; Yamazaki, M.; Maruyama, S. Role of platelet-activating factor in aggregation of
leukocytes and platelets in cerebral ischemia. *Lipids,* **1991**, *26*(12Part2), 1247-1249.
[http://dx.doi.org/10.1007/BF02536541] [PMID: 1819716]
(d) Chung, K.F.; Barnes, P.J. Role for platelet-activating factor in asthma. *Lipids,* **1991**, *26*(12Part2),
1277-1279.
[http://dx.doi.org/10.1007/BF02536547]
(e) Page, C.P. The contribution of platelet-activating factor to allergen-induced eosinophil infiltration
and bronchial hyperresponsiveness. *Lipids,* **1991**, *26*(12Part2), 1280-1282.
[http://dx.doi.org/10.1007/BF02536548] [PMID: 1819716]
(f) Godfroid, J.J.; Dive, G.; Lamotte-Brasseur, J.; Batt, J.P.; Heymans, F. PAF receptor structure: A
hypothesis. *Lipids,* **1991**, *26*(12Part1), 1162-1166.

[http://dx.doi.org/10.1007/BF02536523] [PMID: 1668112]

(g) Lamotte-Brasseur, J.; Heymans, F.; Dive, G.; Lamouri, A.; Batt, J.P.; Redeuilh, C.; Hosford, D.; Braquet, P.; Godfroid, J.J. PAF receptor and "cache-oreilles" effect. Simple PAF antagonists. *Lipids,* **1991**, *26*(12Part1), 1167-1171.
[http://dx.doi.org/10.1007/BF02536524] [PMID: 1668113]

[53]   Chemler, S.R.; Iserloh, U.; Danishefsky, S.J. Enantioselective synthesis of the oxadecalin core of phomactin A *via* a highly stereoselective Diels-Alder reaction. *Org. Lett.,* **2001**, *3*(19), 2949-2951.
[http://dx.doi.org/10.1021/ol0161357] [PMID: 11554815]

[54]   Yao, S.; Johannsen, M.; Audrain, H.; Hazell, R.G.; Jørgensen, K.A. Catalytic Asymmetric Hetero-Diels−Alder Reactions of Ketones: Chemzymatic Reactions. *J. Am. Chem. Soc.,* **1998**, *120*(34), 8599-8605.
[http://dx.doi.org/10.1021/ja981710w]

[55]   Vadhadiya, P.M.; Jean, M.A.; Bouzriba, C.; Tremblay, T.; Lagüe, P.; Fortin, S.; Boukouvalas, J.; Giguère, D. Diversity-Oriented Synthesis of Diol-Based Peptidomimetics as Potential HIV Protease Inhibitors and Antitumor Agents. *ChemBioChem,* **2018**, *19*(16), 1779-1791.
[http://dx.doi.org/10.1002/cbic.201800247]

[56]   Michalik, A.; Hollinshead, J.; Jones, L.; Fleet, G.W.J.; Yu, C.Y.; Hu, X.G.; van Well, R.; Horne, G.; Wilson, F.X.; Kato, A.; Jenkinson, S.F.; Nash, R.J. Steviamine, a new indolizidine alkaloid from Stevia rebaudiana. *Phytochem. Lett.,* **2010**, *3*(3), 136-138.
[http://dx.doi.org/10.1016/j.phytol.2010.04.004]

[57]   Goyal, S.K.; Samsher, ; Goyal, R.K. Stevia (*Stevia rebaudiana*) a bio-sweetener: a review. *Int. J. Food Sci. Nutr.,* **2010**, *61*(1), 1-10.
[http://dx.doi.org/10.3109/09637480903193049] [PMID: 19961353]

[58]   Santhanam, V.; Ramesh, N.G. A Glycal Approach to the Synthesis of Steviamine Analogues. *Eur. J. Org. Chem.,* **2014**, *2014*(31), 6992-6999.
[http://dx.doi.org/10.1002/ejoc.201402943]

[59]   Shan, L.; Sun, P.; Li, T.; Kurtan, T.; Mandi, A.; Antus, S.; Krohn, K.; Draeger, S.; Schulz, B.; Yi, Y.; Ling, L.; Zhang, W. *J. Org. Chem.,* **2011**, *76*, 9699.
[http://dx.doi.org/10.1021/jo201755v] [PMID: 22011230]

[60]   Pazó, M.; Zúñiga, A.; Pérez, M.; Gómez, G.; Fall, Y. Total synthesis of (+)-decytospolides A and B. *Tetrahedron Lett.,* **2015**, *56*(24), 3774-3776.
[http://dx.doi.org/10.1016/j.tetlet.2015.04.055]

[61]   He, H.; Ratnayake, A.S.; Janso, J.E.; He, M.; Yang, H.Y.; Loganzo, F.; Shor, B.; O'Donnell, C.J.; Koehn, F.E. Cytotoxic Spliceostatins from *Burkholderia* sp. and Their Semisynthetic Analogues. *J. Nat. Prod.,* **2014**, *77*(8), 1864-1870.
[http://dx.doi.org/10.1021/np500342m] [PMID: 25098528]

[62]   Nakajima, H.; Hori, Y.; Terano, H.; Okuhara, M.; Manda, T.; Matsumoto, S.; Shimomura, K. New Antitumor Substances, FR901463, FR901464 and FR901465. II. Activities against Experimental Tumors in Mice and Mechanism of Action. *J. Antibiot. (Tokyo),* **1996**, *49*(12), 1204-1211.
[http://dx.doi.org/10.7164/antibiotics.49.1204]

[63]   Nakajima, H.; Takase, S.; Terano, H.; Tanaka, H. New Antitumor Substances, FR901463, FR901464 and FR901465. III. Structures of FR901463, FR901464 and FR901465. *J. Antibiot. (Tokyo),* **1997**, *50*(1), 96-99.
[http://dx.doi.org/10.7164/antibiotics.50.96]

[64]   Motoyoshi, H.; Horigome, M.; Ishigami, K.; Yoshida, T.; Horinouchi, S.; Yoshida, M.; Watanabe, H.; Kitahara, T. Structure-activity relationship for FR901464: a versatile method for the conversion and preparation of biologically active biotinylated probes. *Biosci. Biotechnol. Biochem.,* **2004**, *68*(10), 2178-2182.
[http://dx.doi.org/10.1271/bbb.68.2178] [PMID: 15502365]

[65]   Kaida, D.; Motoyoshi, H.; Tashiro, E.; Nojima, T.; Hagiwara, M.; Ishigami, K.; Watanabe, H.; Kitahara, T.; Yoshida, T.; Nakajima, H.; Tani, T.; Horinouchi, S.; Yoshida, M. Spliceostatin A targets SF3b and inhibits both splicing and nuclear retention of pre-mRNA. *Nat. Chem. Biol.,* **2007,** *3*(9), 576-583.
[http://dx.doi.org/10.1038/nchembio.2007.18] [PMID: 17643111]

[66]   Ghosh, A.K.; Reddy, G.C.; MacRae, A.J.; Jurica, M.S. Enantioselective Synthesis of Spliceostatin G and Evaluation of Bioactivity of Spliceostatin G and Its Methyl Ester. *Org. Lett.,* **2018,** *20*(1), 96-99.
[http://dx.doi.org/10.1021/acs.orglett.7b03456] [PMID: 29218995]

SUBJECT INDEX

A

Acceptors 24, 27, 30, 32, 33, 34, 46, 47, 48, 51, 52, 53, 55, 60, 61
  borate 53
  boron 52
Acetylation reaction 124
Acid(s) 1, 13, 16, 38, 41, 42, 50, 51, 56, 57, 61, 73, 74, 78, 80, 111, 156
  arylboronic 51, 80, 111
  boronic 50
  carboxylic 56, 156
    -catalyzed direct-stereoselective synthesis 42
  Lewis 13, 61, 73, 74
  nucleic 1
  oleic 78
  perfluorophenylboronic 41
  sialic 38
  sugar amino 16
  thiohydroamic 57
Alcohol(s) 2, 32, 35, 37, 38, 41, 43, 45, 53, 111, 115, 127, 145, 149, 163
  aliphatic 38
  derivatives 53
  propargyl 35
  substituted Baylis-Hillman 127
Alkyl halides 12, 88, 89, 90, 91, 92
  dehalogenates 12
Allylmagnesium bromide 153
Amidoglycosylation 48
Amino carbonylation 114
Aminoglycosides 33
Aminooctose moiety 58
Anomeric 26, 27, 29, 30, 41
  effect 27, 29, 30
  nucleophiles 41
  stereoisomers 26
Anomeric carbon 59, 69, 74
  electrophilic 74
Anthracene scaffold 157
Anti-diabetic 14, 160

activity 160
  drugs 14
Anti-fungal activity 146
Anti-inflammatory effects 58
Antibacterial activity 150
Antibiotics 27, 58
  aminoglycoside 27
  drug 27
  thioglycoside-based 58
Anticancer 70, 145
  activity 70
  efficacy 145
  medications 145
Atom-economic glycosylation 62

B

Bacterial protein synthesis, inhibiting 58
Bacteriostatic activity 58
Bioactive properties 125
Biological activities 14, 24, 27, 58, 70, 116, 122, 164
Boranated glucal 146
Buchwald-Hartwig-Migita cross-coupling method 111

C

Cancer treatment 143
Carbine-catalyzed glycosylation 127
Carbohydrate 1, 2, 14, 69, 70, 72, 74, 78, 102, 106, 114, 116, 122, 127, 137
  chemistry 1, 2, 72, 74, 78, 102, 106, 114, 116, 122, 127
  derivatives 2, 69, 70, 137
  moiety 14
  scaffold 1, 2
Carbon monoxide 114
Carbonylative esterification 115
Cardiac glycoside 27
  lipid-soluble 27
Catalysis 60, 72, 97, 120

cobalt 60
photo-redox 72, 97, 120
Catalyst 30, 35, 39, 45, 47, 52, 60, 94, 99,
      101, 112, 116, 117, 120, 124, 155
  -catalyzed olefin metathesis 155
  -free technique 99
  -mediated reaction 52
Catalytic 37, 59, 77, 83
  cycle 77, 83
  system 37, 59
Catalyzed stereoselective glycosylation 51
Chemical 1, 26, 74, 114
  industries 114
  synthesis 1, 26
  waste 74
Chemistry, organic 1, 21, 116, 137, 138
Chiral building blocks 1, 108, 141
Chloride 33, 49, 50, 140, 142
  gold 33
  -mediated glycosylation 49
  oxalyl 142
  sodium 140
Chromium, reactive 9
Chromone derivatives 79
Coupling 88, 92, 117, 146, 161
  process 146
  reaction 88, 92, 117, 161
Cross-coupling carbonylative 109
Cross-coupling reactions 109, 111, 128
  ligand-free Suzuki-Miyaura 111
  metal-catalyzed 109
  palladium-based 128
Cycloaddition 122, 158
  process 158
  reaction 122
Cycloalkenones 116
Cycloalkenons 116
Cyclohexanone 140
Cyclopropanated 127, 130, 131, 132
  glycals 131, 132
  sugars 127, 130
Cyclopropanation reaction of glycals 132
Cyclopropane 127, 128, 129, 131, 145
  glycal-derived 128
  halogenated 128
Cytotoxic activity 141
Cytotoxicity 150, 153

D

Deacetylation, bicatalytic 159
Debenzylation, palladium-catalyzed 149
Decarboxylative 92, 155
  coupling product 155
  glycosylation 92
Dehydrative reaction condition 109
Deoxyglycosides 37
Deprotection 53, 140, 146, 150, 153, 157
  acetonide 140
Deprotonation 86
Desulfoglucosinolates 57
Diastereomeric combination 143
Diastereoselective cyclopropanation reaction
  132
Dicotyledonous angiosperms 58
Diels-Alder cycloaddition 116, 124, 158
  reaction 124
Diels-Alder reaction 159
Diospongin, steroid 138
Dipole-dipole interaction 29, 30
Disaccharides 1, 30, 32, 36, 41, 44, 45, 46,
      109, 112, 122, 128
  formed 128
Drug 27, 69, 102, 138
  anticancer 27
  development 102, 138
  discovery programs 69

E

Edward-Lemieux effect 29
Electrochemical reduction 10
Electrolysis 10
Electron 19, 119
  -deficient alkene sources 119
  -donating nature 19
Electrophilic attacks 107
Elimination, reductive 4, 77, 89, 94
Empagliflozin 14, 69, 138
Enzymatic degradation 69
Erythromycin 27
Ester 4, 5, 26, 32, 94, 109, 110
  bond 26
  protections 4
  sulfamate 32
Esterification 142
Ethyl isocyanoacetate 15, 16, 17

F

Facile glycosylation 35
Ferrier 33, 46, 110
  compounds 110
  glycosylation 33, 46
  reaction 110
Ferrier rearrangement 13, 34, 43, 69, 73, 74, 107
  controlled 43
Fischer-Zach method 6
Fluorescence analysis 56
Functions, biological 138
Fungus 150, 162
  endophytic 162
Furan-based enol ethers 119
Furanose moieties 1
Furo, glycal-based 124
Furonoid glycals 1

G

Galacotosides 85
Galactal 2, 33, 36, 46, 57, 84, 86, 88, 90, 100
  acetylated 36
  -based iodo-pyranone 100
  benzylated 90
  benzyl-protected 46
Galactosides 33, 39, 40, 46, 84
Galactosylsulfoxide 6
Generation, oligosaccharide 24
Glucal 12, 33, 84, 87, 88, 92, 99, 114, 120, 141, 142, 145, 146, 149, 151, 155, 159, 160, 161, 162, 163
  acetylated 163
  acyloxylated 120
  anomeric-activated 146
  -based iodo-pyranone 99
  benzylated 92, 160
  benzyl-protected 88
  ester-protected 114
  product, desired 12
  protected 141, 155
  reacting 33
Glucopyranoside moiety 146
Glucosinolates 24, 57, 58
Glucosyl, anticipated 84
Glycal 3, 4, 6, 7, 8, 12, 13, 25, 32, 36, 38, 39, 45, 48, 51, 76, 81, 90, 97, 98, 106, 109, 119, 122, 132

acetylated 32
acetyl-protected 36, 48, 90, 122
acylated 36
derivatives, protected 76
donors 38, 39, 45, 48, 51, 97, 98
ester-protected 4
furanoid 7, 8, 12, 109
furanose 3
moiety 106, 119
pyranoid 6, 7, 12
structures 81
transformation of 4, 25
transformations 13
transforming 132
Glycoconjugates 30, 69, 125
Glycoproteins 1
Glycosidase 122, 125
  inhibitors 122
  inhibitory activities 125
Glycosidations 1
Glycosides 27, 39, 41, 44, 45, 56, 146
  desired 27, 39, 41, 44, 45, 56
  matching 146
Glycosyl 4, 6, 7, 10, 11, 12, 39, 70, 97
  halides 10, 12, 70
  halides and glycals 11
  radicals 97
  sulfoxides 4, 6, 7
  transferases enzyme 39
Glycosylated products 78
Glycosylation 24, 26, 36, 38, 41, 43, 44, 69, 70, 84, 95
Glycosylation 25, 26, 53, 55, 73
  conditions 26
  of glycals 25, 55, 73
  process 25, 53
Glycosylation reactions 53, 72, 90, 107
  ligand-controlled stereoselective C-alkyl 90
Grubbs catalyst 152, 156
Grubb's cyclization 156

H

Halogen atoms 120
Halogenated cyclopropane glucals 128, 130
Heck reaction 110
Hemiacetal moiety 142
Hydrogenation 4, 16, 107, 124, 143, 152, 162, 163
  catalytic 16

palladium-catalyzed 152
Hydrogenolysis 162
Hyperconjugation 29
Hypoglycemia 48
Hypotension 27, 157

I

Infections 27, 149
  bacterial 27
  respiratory tract 27
  skin 27
  viral 149
Iodoamination 154
Iridium-promoted deoxyglycoside synthesis
  48
Isomeric product 19
Isotope labeling 50

L

Lactol, produced 140
Lactones 17, 160
  heating 160
  sugar-derived 17
Landomycin 27
LDA conditions 150
Ligand(s) 38, 75, 77, 112, 114, 117
  bidentate 38
  exchange 77
Limitation-induced cell death 49
Lincomycin 58
Linkage, glycosidic 24, 25
Lipopolysaccharides 1, 140
Luche's conditions 158

M

Macrolactone 152
Macrolide, produced 142
Martin periodinane oxidation 152
Mechanism 12, 26, 27, 50, 132
  radical-based 12
Mechanistic pathways 26, 29, 73, 77, 94
Medicinal applications 28
Medicines, traditional 116
Menomycin 27
Mercury cathode 10
Metal-free synthesis 99
Method, electrochemical 1, 4, 10, 11

Micrococcus luteus 150
Microwave irradiation 20, 98
Mitsunobu reaction 152
Miyaura cross-coupling reaction 163
Monosaccharides 1, 25, 112

N

Natural 2, 108, 120, 138, 162
  molecules 138
  product synthesis 2
  products, glycal-derived 108, 120
  products decytospolide 162
Nature, electronic 2
Nitration-scission reaction 126
Nitro-polyol derivatives 125
Nitrogen 59, 153
  atmosphere 59
  heterocycles 153
NMR spectroscopy 101
Nucleophiles, aliphatic 46
Nucleophilic 57, 59, 74, 107, 159
  allenic 74
  substitution 159
  sulfur 59

O

Oligosaccharides 24, 30, 40, 53, 61
One-pot 55, 144
  hydrozirconation/transmetallation method
    144
  reaction 55
Organic synthesis 14, 138
Organocatalyst, effective 74
Organometallic reagents 10
Oxidative 40, 41
  aerobic glycosylation 40
  glycosylation 41
Oxocarbenium ion 27, 29, 30, 48, 74
Oxygen 51, 57, 69
  labile glycosidic 69
  source 51
Ozonolysis 4, 107

P

Palladium 38
  catalysis 38
  catalyst 38

Pd-catalyzed 18, 59, 60, 84, 85, 86, 87, 117, 128, 157
   allylic rearrangement 59
   approach 60, 84
   coupling 18, 128, 157
   decarboxylative reaction 85
   Hiyama reaction 86, 87
   synthesis 84, 117
Peptidoglycans 1
Peptidomimetics 159, 160
Peracetylated 12, 32
   glycosyl bromide 12
   glucals react 32
Petasis reagent 145
Phomactin 157, 158, 159
Phosphonoselenide catalyst 101
Photocatalyst 120
Photoredox catalysis 69
Platelet 157, 158
   -activating factor (PAF) 157, 158
   -activating factor-mediated signaling 158
   aggregation 157
Plausible reaction pathway 83
Precursors, biosynthetic 1
Production, stereoselective 96
Propargylic carboxylates 73, 74
Properties 14, 58, 150, 163
   biological 14
   cancer chemoprotective 58
   cytotoxic 163
   robust anti-inflammatory 150
Protection 4, 82, 88, 140, 149, 156, 163
   allylic 4
   -deprotection 149
Proteoglycans 1
Protocol 25, 30, 38, 40, 43, 45, 46, 78, 84, 98, 101, 102
   metal-free 102
Pseudoglycals 119
Pulmonary disorders 158
Pyridinium chlorochromate 153
Pyrimidinone glycohybrids 99, 100

R

Reactivity of glycals 13, 107
Redox-active esters (RAE) 92
Regioselective 39, 40, 44, 50
   acid-catalyzed 40
   glycosidations 50

glycosylation 39, 44
Reticulated vitreous carbon (RVC) 10
Retro-Henry-type reaction 125

S

Selective glycosylation 43
Spliceosome inhibition 145
Stability, metabolic 69
Stereochemical 6, 59
   control 59
   variations 6
Stereoelectronic effect 29
Stereoisomer, single 124
Stereoselective 37, 41, 45, 47, 55, 59, 74, 77, 78, 79, 92, 126, 128, 129, 138, 153, 158
   coupling reaction 92
   glycosylations 74
   one-pot method 126
   one-pot synthesis 55
   reactions 138
   synthesis 37, 45, 47, 77, 78, 79, 129, 158
   thioglycosylation 59
Streptomyces lincolnensis 58
Streptomycin 27
Sugar(s) 129, 132
   acetylated 129
   -derived cyclopropanes 132
Sugar derivatives 4, 21, 34, 70
   electrophilic 70
Sulfoxide, galactose-derived 6
Suzuki coupling 145
Synthesis 10, 25, 61, 102, 147, 148, 161, 162, 163, 164
   aspicilin 161
   of glycals synthesis 10
   of natural products decytospolide 162
   of Papulacandin 147, 148
   of spliceostatin 163, 164
   oligosaccharide 25, 61
   one-pot 61, 102

T

TEOC protection 146
Transition metal catalysts 114

Y

Yamaguchi's conditions 152

www.ingramcontent.com/pod-product-compliance
Lightning Source LLC
Chambersburg PA
CBHW041704210326
41598CB00007B/523